Made in New York

25 Innovators Who Shaped Our World

FRANK VIZARD

EXCELSIOR
EDITIONS

Cover images, L to R: "Kodak Girl": Woman holding early Kodak Camera.
Library of Congress, Prints and Photographs Division; The U.S.S. Plunger, c. 1905.
Library of Congress Prints and Photographs Division, Bain News Service;
Nicola Tesla, c. 1890. Library of Congress Prints and Photographs Division,
Bain News Service.

Published by State University of New York Press

Excelsior Editions is an imprint of State University of New York Press

For information, contact State University of New York Press, Albany, NY
www.sunypress.edu

Library of Congress Cataloging-in-Publication Data

Name: Vizard, Frank, author.
Title: Made in New York : 25 innovators who shaped our world / Frank Vizard.
Other titles: Made in New York, twenty-five innovators who shaped our world
Description: Albany : State University of New York Press, 2023. | Series:
    Excelsior editions | Includes bibliographical references and index.
Identifiers: LCCN 2023002663 | ISBN 9781438493688 (pbk. : alk. paper) |
    ISBN 9781438493671 (ebook)
Subjects: LCSH: Inventions—New York (State)—History. | Inventors—New
    York (State)—Biography. | New York (State)—Civilization—Miscellanea. |
    Inventors—New York (State)—New York—Biography. | New York (N.Y.)—
    Civilization—Miscellanea.
Classification: LCC T22.N7 V59 2023 | DDC 609.2/2747—dc23/eng/20230228
LC record available at https://lccn.loc.gov/2023002663

10 9 8 7 6 5 4 3 2 1

The present is theirs; the future, for which I really worked, is mine.

—Nikola Tesla

For my daughter, Nuala.

Contents

List of Illustrations xi

Introduction: Made in New York 1

Chapter 1   The Dry Cleaner and the Abolitionist (1821) 5

Chapter 2   A Diamond Is Forever: Devising the Rules of
            Baseball (1845) 11

Chapter 3   The High-Class Origin of the Potato Chip (1850) ... 17

Chapter 4   Going Up: The Elevator Makes New York
            Skyscrapers Possible (1854) 22

Chapter 5   The Submarine Paid For by Irish Rebels (1881) 28

Chapter 6   The First Roller Coaster (1884) 35

Chapter 7   Photography's Kodak Moment (1887) 41

Chapter 8   The Machinery of Democracy (1892) 48

Chapter 9   Tesla Invents the Remote Control (1898) 55

Chapter 10 The Cool Machine (1902) 64

Chapter 11 The Teddy Bear Is Born in Brooklyn (1902) 71

Chapter 12   The Teenage Debutante Who Let Women
              Breathe Easier (1910)   77

Chapter 13   The Man Who Streamlined America (1934)   84

Chapter 14   *Scrabble* Games the World (1938)   91

Chapter 15   The Invisible Woman Makes *Gone with the Wind*
              a Hit Movie (1939)   97

Chapter 16   Batman Is Born in the Bronx (1939)   102

Chapter 17   The Bloodmobile Drives to War (1940)   109

Chapter 18   The Atomic Woman (1945)   115

Chapter 19   The LP Makes Vinyl a Hit Record (1948)   122

Chapter 20   How to Mend a Broken Heart (1960)   128

Chapter 21   New York Traffic Inspires the Maglev Train (1966)   134

Chapter 22   A Queens Nurse Invents the Security Camera
              (1966)   141

Chapter 23   Break This: The Birth of Hip-Hop (1973)   145

Chapter 24   The Body Scanner (1977)   151

Chapter 25   Solar Lights Up the Dangerous Dark (2015)   158

Index   163

Illustrations

Figure 1   A portrait of a man alleged to be Thomas L. Jennings.   5

Figure 2   Alexander Joy Cartwright wearing a fireman's helmet.   11

Figure 3   Moon's Lake House, Saratoga Springs.   17

Figure 4   Equitable Life Insurance Building showing elevator banks on left.   22

Figure 5   The USS *Plunger*, c. 1905.   8

Figure 6   Scenic Railway entrance, Luna Park, 1906.   35

Figure 7   "Kodak Girl": Woman holding early Kodak camera.   41

Figure 8   Patent drawing for J. H. Myers's voting machine.   48

Figure 9   Nikola Tesla, c. 1890.   55

Figure 10   Willis Carrier, 1915.   64

Figure 11   Teddy bears made in New York City.   71

Figure 12   Phelps's 1914 patent drawing for her brassiere.   77

Figure 13   Raymond Loewy with a model of his Imperial House II.   84

Figure 14   *Scrabble* inventor Alfred Butts and promoter James Brunot playing on a large *Scrabble* board.   91

Figure 15   Katharine Blodgett demonstrating her invention, 1938.   97

Figure 16   *Batman* Issue no. 1.   102

Figure 17   The Howard University medical unit headed by Dr. Charles Drew treating a patient from the Bloodmobile in a practice drill, 1943.   109

Figure 18   Chien-Shiung Wu, 1958.   115

Figure 19   Labels for Columbia Long-Playing Records.   122

Figure 20   Wilson Greatbatch holding the original pacemaker.   128

Figure 21   A maglev train coming out of the Pudong International Airport.   134

Figure 22   Patent drawing for Marie Van Brittan Brown's home security system.   141

Figure 23   Afrika Bambaataa.   145

Figure 24   MRI image of a brain.   151

Figure 25   SolarPuff lantern.   158

Introduction

Made in New York

When Frank Sinatra sang "New York, New York," perhaps the legendary crooner's most famous song, he proclaimed "If I can make it there, I'll make it anywhere." What Sinatra neglected to mention was all the things invented in New York that went everywhere around the globe. Throughout its history, New York has been the epicenter of life-changing invention.

New York's inventive influence in almost every human endeavor is outsized. Every school kid learns how Robert Fulton invented the steamboat and sailed it up the Hudson River in 1807, ushering in a new age of transportation. With the opening of the Erie Canal in 1825, New York became the most important center for commerce in North America. As New York grew, so did the inventive minds of its population, one that was determined to leave a mark.

That mark could range from the mundane to the grandiose. Seth Wheeler of Albany may be a name largely lost to history but his invention of perforated toilet paper in 1871 is used daily. Billiard balls developed by another Albany inventor, John Wesley Hyatt, in 1863 still carom around modern pool tables. The widely used, porch-loving Adirondack chair is a creation of Thomas Lee and Harry Bunnel, who lived in Westport by Lake Champlain at the turn of the twentieth century.

On a grander scale, Long Islander Leroy Grumman co-founded an aviation company in 1929 that has since morphed into the giant defense contractor Northrup Grumman. Likewise, George Eastman of Syracuse made Kodak synonymous with photography. Isaac Singer of Pittstown turned sewing machines bearing his name into one of the first multinational companies. Of course, Willis Carrier of Angola is the coolest, forever linked to the air conditioners that still bear his name.

Other inventive geniuses labored under the umbrella of corporations. At General Electric, Charles Proteus Steinmetz, a man who stood four-feet tall, was a giant in the field of electromagnetism and electric motors during the start of the twentieth century, earning the nicknames "Wizard of Schenectady" and "Forger of Thunderbolts." In 1913, a Corning physicist named Dr. Jesse Littleton invented a heat-resistant glass called Pyrex, which became a chef's best friend. New York became home to numerous corporate research laboratories belonging to major companies like the computer maker IBM and national labs like physics-oriented Brookhaven, all of which made their own innovative contributions.

Like many New Yorkers, some inventors came from elsewhere but had their greatest successes developing innovations in their adopted home—a roster that would include superstar luminaries such as Nikola Tesla and Raymond Loewy. In some instances, inventors developed their creations elsewhere but eventually became New York residents. Madam C. J. Walker, the developer of hair-care products for Blacks, was perhaps the world's wealthiest woman when she died in Irvington, New York, at the age of fifty-one in 1919. Conversely, some inventors born and educated in New York made their mark once they left. Harlem-born and New York educated Dr. Patricia Bath was the first Black woman physician to obtain a patent and was an inductee into the 2022 National Inventors Hall of Fame for a 1988 medical device that removed cataracts—invented after she moved to California to become the first woman ophthalmologist at UCLA's Jules Stein Eye Institute. Some inventors who never lived in New York made their mark from afar, like little-known Anna Connelly of Philadelphia whose lifesaving exterior fire escape, invented in 1887, became an iconic feature of New York architecture.

New Yorkers are creative inventors when they're hungry. Despite their names, the English muffin, the Baked Alaska, General Tso's chicken, pasta primavera, and Napoleon cookies were creations of New York chefs. Add the Reuben sandwich, eggs Benedict, red velvet cake, the Waldorf salad, hamburgers, hot dogs, Oreo cookies, and the potato chip to the New York menu. A New York restaurant could probably get by with just serving New York–created food.

Cocktails like the manhattan and the cosmopolitan were born out of New York nightlife as was drinking a bloody mary the morning after, although drink origin stories are often a little fuzzy. New York nightlife also gave birth to or showcased nearly every form of music. Two of Manhattan's most famous music venues, the Metropolitan Opera House and Carnegie Hall, were constructed in the nineteenth century. Broadway is the home address of musical theater. Venues large and small like Studio 54, CBGB, the Village Vanguard, the Bottom Line, the Cotton Club, the Blue Note, Max's Kansas City, the Palladium, and Madison Square Garden are legendary. Bebop jazz, salsa, punk rock, and hip-hop were born out of New York's cultural diversity. In New York, there's always an ear for a new song.

Any list of New York inventions is a formidable one that cuts across every endeavor. MADE IN NEW YORK tells the stories of New York inventions that have had a significant impact on the world stage. Perhaps more to the point, their origin stories are surprising relevant in modern times. New Yorkers are no strangers to adversity. Many New York inventions were born out of adverse conditions and their inventors often faced challenging circumstances. Even something seemingly as innocent as the Brooklyn-born teddy bear has a backstory of racism. Some inventors never benefitted financially from the fruits of their labor. Others waited decades for recognition. Some are still relatively unknown outside a small circle of insiders in their respective fields.

Most were genius visionaries and risk-takers way ahead of their time. Many, of course, succeeded beyond their wildest dreams and were amply rewarded for their perseverance, often channeling their new-found wealth toward worthy causes because that's what this breed of New Yorkers are about. It would be fair to call all of them heroes. Their stories are connected in that their innovations have become cornerstones of modern life.

Sinatra had it right.

1

The Dry Cleaner
and the Abolitionist (1821)

Figure 1. A portrait of a man alleged to be Thomas L. Jennings. *Source*: Public domain.

Dry cleaning and civil rights aren't two subjects that would seem intrinsically linked but for Thomas L. Jennings, the Black man who invented the dry-cleaning process, the two are inexorably connected. Jennings was the first Black man to receive a patent from the United States. That patent would help fund a civil rights movement in the years prior to the Civil War. The famous Black orator and abolitionist Frederick Douglass called Jennings "a bold man of color" in a eulogy to mark his passing.

Born in 1791, Jennings was a free-born New Yorker during a period when slavery for African-Americans was the norm. He also was an avowed patriot. During the War of 1812, Jennings was among "the one thousand citizens of color" who volunteered to dig trenches to fortify New York against a British attack.

Jennings apprenticed as a tailor and then opened his own shop at the age of nineteen on Church Street in Manhattan. Jennings's skill as a tailor became widely appreciated and he expanded his business to become a clothier. Inspired by the many complaints of customers whose clothes had become irrevocably soiled, Jennings began experimenting with a variety of cleansing chemicals. Jennings finally hit on a process he would call "dry scouring" that would clean clothing without damage from water soaking through the fibers. The exact methodology Jennings employed has been lost to history as it became one of the ten thousand "X" patents destroyed in a fire in 1836 at a Washington, DC, hotel where patent records had been temporarily stored. When a rival tailor used Jennings's widely recognized process, Jennings sued him and won, dramatically producing what's now called patent 3306x.

The patent grant was extraordinary for its time. United States patent laws written in 1793 stated that "the master is the owner of the fruits of the labor of the slave both manual and intellectual." This meant slaves could not patent or profit from their own inventions. Jennings, however, was a free man so, as such, he gained exclusive rights to his process. And as Douglass would later note, the patent recognized Jennings as a citizen of the United States, an official acknowledgement that caused a stir in many pro-slavery circles. Congress would extend patent rights to slaves in 1861 at the start of the Civil War.

Jennings's dry scouring process made him a prosperous man. Jennings's first move with his newfound wealth was an act of devotion. Jennings fell in love with a woman who had been born into slavery and tried to buy her freedom. However, New York State laws made emancipation a drawn-out process designed to protect slave owners from financial loss. His wife Elizabeth's status was converted to "indentured servant," and she was not fully emancipated until 1827. They would have four children together. Jennings would spend much of his earnings on legal fees to buy his children out of bondage. Before 1827, children born to slave mothers were considered free but were required to serve years-long apprenticeships to their mother's master. Even when freed, Blacks continued to look over their shoulder as a "New York Kidnapping Club" would snatch people off New York City streets for transportation and sale south, crimes often ignored by the authorities, notes author Jonathan D. Wells in his 2020 book *The Kidnapping Club: Wall Street, Slavery, and Resistance on the Eve of the Civil War*. The kidnapping danger was long-standing. One well-publicized rescue of nine kidnapped freed Blacks, ranging in age from eight to twenty-three, from a ship called the *Creole* anchored near Ossining, was reported by the *New York Evening Post* in June 1817.

It's what Jennings did next that left its mark on history. Jennings used his fortune to become the money man for a variety of nascent abolitionist movements all over the northeast United States. In 1831, Jennings was named the assistant secretary for the First Annual Convention of the People of Color in Philadelphia, a meeting the presaged the founding of the National Association for the Advancement of Colored People (NAACP) in 1909.

The Philadelphia convention and subsequent ones that became collectively known the Black Convention Movement were the scenes of intense debate. Delegates were free or formerly enslaved African-Americans from a broad spectrum of religious leaders, businessmen, politicians, writers, and editors dedicated to the cause of abolitionism long before any formal anti-slavery movements began in the United States.

Perhaps the most contentious issue involved the American Colonization Society (ACS) founded by Robert Finlay, a Presbyterian minister from New Jersey. The ACS proposed sending free Blacks to a colony in Liberia, Africa, as a gradual end to slavery. The membership of the ACS was overwhelmingly white, all of whom agreed that free Blacks could not be successfully integrated into American society. The ACS received some initial funding from the federal government, which helped with the purchase of land in Liberia, although most funding came from private sources. The ACS's influence was great enough to transport thousands to the West African country over the course of the organization's existence. Many Blacks were sympathetic to the ACS notion that their full potential could only be realized by returning to Africa.

Jennings adopted what was then a minority view, slamming the ACS mission. "Our claims are on America; it is the land that gave us birth; it is the land of our nativity, we know no other country, it is a land in which our fathers have suffered and toiled; they have watered it with their tears, and fanned it with sighs," said Jennings in a speech. "Our relation with Africa is the same as the white man's with Europe, only with this one small difference, the one emigrated voluntarily, the other was forced from home and all its pleasures."

Jennings put his money where his mouth was, funding *The Freeman's Journal*, the first weekly newspaper aimed at African-Americans. The New York publication opposed colonization and promoted Black suffrage and education. Gradually, the majority of opinions came around to Jennings's viewpoint. In 1847, Liberia declared its independence from the ACS.

Jennings used his wealth to fund various abolitionist causes. He was a founder and trustee of the Abyssinian Baptist Church and backed other philanthropic organizations as well. His children followed in his footsteps, with one son becoming a leading abolitionist in Boston. Another son worked on anti-slavery committees with Frederick Douglass. Jennings's wife and two daughters were active in the Female Literary Society of New York, which raised money to free slaves and promoted the rights of African-American women.

Jennings also funded many legal challenges against discrimination, but the most significant case involved his youngest daughter, who became Rosa Parks long before there was the Rosa Parks who famously refused to give up her seat on a bus in 1955 in Montgomery, Alabama.

In June of 1854, Elizabeth Jennings, the youngest and only child born free, was on her way to church and boarded the Third Avenue Railroad, a horse-drawn network of streetcars serving Manhattan, at the corner of Pearl and Chatham Streets. As recounted by her in a first-person account published in the anti-slavery newspaper the *New-York Daily Tribune*, the conductor told the twenty-four-year-old schoolteacher to wait for the next car as it would be accepting Black people. Jennings, ever her father's daughter and named after her mother, argued with the conductor. "I have no people," she wrote. "I wished to go to church."

After brief stalemate, the conductor agreed to let Jennings on the streetcar but that she would have to disembark if any white people boarded. Jennings would have none of it. "I told him I was a respectable person, born and raised in New York and that he was a good-for-nothing impudent fellow for insulting decent persons on their way to church."

The confrontation turned physical when the conductor, with the help of the driver, tried to forcefully remove Jennings. "I screamed murder with all my voice," said Jennings, who remained successfully aboard, albeit on the floor of the streetcar "so that my feet hung one way and my head the other, nearly on the ground." Her hat was ruined, and her dress torn. The conductor instructed the driver to head for the nearest police station where an officer helped in her removal, subsequently taunting her "to get redress if she could."

She could. And a future United States president helped.

Representing Elizabeth was the law firm hired by her father, Culver, Parker & Arthur, which was already involved in legal work on behalf of abolitionists. Chester Arthur, who would become the twenty-first president of the United States in 1881, was at that time a fresh-faced lawyer with unruly hair and immense sideburns who had only passed the bar two months before, becoming the firm's newest partner. The entire African-American community in New York was livid. Frederick Douglass noted that Elizabeth's African-born grandfather, Jacob Cartwright, had been a soldier in the Revolutionary War. Arthur, the son of an abolitionist preacher, won the case before an all-white male jury in a New York State court convened in Brooklyn. Slavery had ended in New York State in 1827 but the Jennings case etched equal treatment into the law. Jennings helped form a legal aid association to represent minorities fighting discrimination. It would be another decade before privately-owned streetcar companies stopped discriminatory practices. Elizabeth would go on to found the city's first kindergarten for Black children.

Thomas Jennings died in 1859. History will remember him for being the first Black man to receive a patent. Jennings's "dry-scouring" method is essentially the one used by dry-cleaning businesses today. But Jennings's real legacy is that his invention helped spur a movement for equality. The epitaph on his headstone reads "Defender of Human Rights."

2

A Diamond Is Forever

Devising the Rules of Baseball (1845)

Figure 2. Alexander Joy Cartwright wearing a fireman's helmet. *Source*: Courtesy Hawaii State Archives.

Baseball was a lawless game in 1845. The sport that would become known as America's national pastime had already been played for many years. The term *base ball* is mentioned in Pittsfield, Massachusetts, in 1791, the word used by an Anglican bishop to decry the playing of the game on Sundays. However, the rules governing baseball were haphazard at best. In some instances, the rules were more akin to the British game of rounders, from which baseball is descended. Other variations required fielders to hit the runner with the ball. And as there was no rule regarding base paths, this led to merry chases into the outfield. The number of players on each team was variable as was the number of bases and the distance between them. The man who brought order out of chaos was a gregarious, strapping New Yorker named Alexander Joy Cartwright. While the development of baseball as we know it today was probably a communal affair that involved many contributions, Cartwright's role is so significant that many, including the National Baseball Hall of Fame, consider Cartwright the father of modern baseball.

Cartwright was the son of merchant sea captain. At the age of sixteen, Cartwright became a bank clerk to help support his large family—Cartwright had six siblings. Bankers' hours meant that Cartwright and other young men involved in the legal and merchant trade often had late afternoons free to engage in sports. Cartwright also volunteered as a fireman, first with Oceana Hose Company No. 36 and later with Knickerbocker Engine Company No. 12. Six feet tall and weighing 215 pounds, Cartwright was a standout player for a team known as the New York Gothams (no record exists of whether he was a lefty or righty). But in 1842, when he was twenty-two and newly married, Cartwright and others broke away to form their own team, named the Knickerbocker Base Ball Club (named after the fire engine company).

Then as now, finding a large space to play baseball in New York City was a challenge. Most games were played on a vacant lot at Twenty-Seventh Street and Fourth Avenue and later at Thirty-Fourth Street and Lexington Avenue. By 1845, these spots were no longer available. Cartwright, now vice president and secretary of the Knickerbockers, and his friends, many of whom were also volunteer firefighters, formally organized the Knickerbockers with a constitution and bylaws to engage in the search for a new regular playing field. A membership fee would help raise the money to secure a field. Club members were divided into various teams. Meanwhile, the bank where Cartwright worked somewhat ironically burned down and Cartwright was in the midst of learning a new trade as a bookseller on Wall Street.

With New York City growing by leaps and bounds, Cartwright's search for a home field finally led him across the Hudson River to Hoboken, New Jersey, to a former cricket field (Elysian Fields), which was available for a seventy-five-dollar yearly fee. But even here, the space was tight with a glue factory to one side and a couple of roads infringing on the space available. The shape of the Elysian Fields dictated the look of the modern game. Cartwright devised a diamond configuration with the batters facing the outfield so that the glue factory and one of the roads was not a problem for players. The second road was now in the far outfield so that was not a problem either. And again, the shape of the Elysian Fields dictated another long-lasting rule—ninety feet between the bases.

Cartwright insisted on playing by a formal set of rules that became known as "the Cartwright Rules" and later the "New York game." The rules were twenty in number and it's likely that many, such as the definition of fair and foul territory beyond the first and third base lines, were informally in use before Cartwright codified them. Three strikes per batter and three outs per inning for each side were in the rule book. Cartwright's major innovations included a nine-inning game and a nine-player team in an unalterable batting order. Clubs eventually established their "First Nine" as well as a "Second Nine" of aspirants with games between the latter often being more popular than the former. Cartwright also made the game safer, mandating that runners could be tagged out rather than being "soaked" by the ball thrown at them, ultimately leading to the use of a heavier ball than what was used at the time. This last change gave the game more dignity while also speeding up the proceedings. Not every rule made it into the modern game—a batted ball caught on one bounce was an out, for example.

Not everyone who had been a member of the Knicks was happy at the move to Hoboken. A splinter group formed the New York Nine but nonetheless became the first opponents to play against the Knicks at Elysian Fields on June 19, 1846. The New York Nine won the game by a score of twenty-three to one. At the time, the first team to reach twenty-one was declared the winner so it's likely that the game ended with a walk-off home run. The score suggests that most of the better players weren't keen on New Jersey as a home base. Cartwright is said to have umpired the game and fined the New York Nine pitcher six cents for cursing, perhaps the first recorded instance of a player tangling with an umpire. Among the spectators in 1847 was Henry Chadwick, a New York journalist who would later develop the box score. The Knickerbockers continued to play at Elysian Fields until 1870.

Cartwright stayed with the Knickerbockers until 1849 just as the Knickerbockers adopted a formal uniform reminiscent of earlier cricket teams: long blue woolen trousers, white flannel shirts, and straw hats. The promise of riches in the California Gold Rush, however, proved impossible to resist. Some accounts portray Cartwright as a Johnny Appleseed of the game, spreading the New York game across the country as he and a dozen friends traveled west, first by train and then by covered wagon, "during which time they managed to work in a little baseball at various stopovers," writes Cartwright in his diary. Newcomers to the game, including Native Americans, didn't play it with the grace of the New Yorkers. "It is comical to see the mountain men (trappers used for guides) and Indians playing the new game," Cartwright wrote. "I have the ball with me that we used back home."

Cartwright reached California after a months-long journey but didn't stay long. Perhaps sensing better opportunities, Cartwright set sail for Hawaii, then known as the Sandwich Islands. Cartwright prospered in Hawaii—his wife and five children would arrive in 1851—and would spend the rest of his life there, becoming a close friend of King Kamehameha IV. Naturally, Cartwright is credited with bringing baseball to Hawaii, firmly establishing the sport in Honolulu even before it gained ground in cities like Detroit and Chicago. Cartwright died in Hawaii on July 12, 1892.

The Cartwright story was virtually unknown and only surfaced as a counter to the claim that Abner Doubleday, best known as the Civil War general who fired the first shot for the Union at Fort Sumter, North

Carolina, was the father of baseball. The Doubleday claim was perpetuated by Albert Spalding, a pioneer team executive with the Chicago Cubs and sporting goods magnate, who took umbrage at the suggestion made in a 1903 newspaper article by Henry Chadwick, by now an acclaimed sportswriter, that baseball was simply a refinement of the old British game of rounders. Spalding went to great lengths to prove baseball was a product of American exceptionalism, instigating the formation of the Mills Commission, named for National League president Abraham G. Mills, in 1905 to investigate baseball's origins. The commission's finding was published in 1908 in the *Spalding Guide*. Baseball had been invented by Doubleday in 1839 in Cooperstown, New York, a location that would become the site of the National Baseball Hall of Fame in 1939. Doubleday, the commission found, had personally modified "town ball," a game that might involve twenty to fifty boys trying to catch a ball hit by a "tosser" with a flat bat.

The Doubleday story came under scrutiny when the Baseball Hall of Fame opened in Cooperstown to mark the alleged centennial of the sport. Abner Graves, the only witness to the 1839 game, whose testimony was the foundation to the Doubleday claim, would have been five years old at the time. Doubleday was a cadet at the military academy at West Point and there is no evidence to suggest he had any interest in the game. Graves, who harbored anti-British sentiments, later killed his wife, was declared insane, and was committed to a psychiatric hospital.

The close scrutiny of the Doubleday claim was also likely prompted by evidence of a Cartwright's pioneering role presented the previous year to the National Baseball Hall of Fame commission by Cartwright's grandson, Bruce Cartwright, Jr. That documentation was accepted by the commission. Alexander Joy Cartwright was inducted into the National Baseball Hall of Fame with a plaque recognizing him as the father of modern baseball. Cartwright was credited with the development of the nine-inning game played with nine players on a team with bases ninety feet apart. August 26, 1939, was declared "National Cartwright Day." Among the celebrants were ballplayers at Ebbets Field in Brooklyn where they drank pineapple juice in a toast to Cartwright. The game that day became the first major league baseball game to be broadcast on television. Henry Chadwick was inducted into the National Baseball Hall of Fame as well.

Doubleday, meanwhile, was never inducted into the National Baseball Hall of Fame. Some writers still question the validity of Cartwright's designation as baseball's patriarch. Indeed, the rules of the game probably evolved over time thanks to the input of numerous influencers. But Cartwright's plaque remains on the wall.

3

The High-Class Origin
of the Potato Chip (1850)

Figure 3. Moon's Lake House, Saratoga Springs. *Source*: The Miriam and Ira D. Wallach Division of Art, Prints, and Photographs: Photography Collection, The New York Public Library, *Moon's Lake House, Saratoga Lake*, New York Public Library Digital Collections, accessed August 16, 2022, https://digitalcollections.nypl.org/items/510d47e1-5bbb-a3d9-e040-e00a18064a99.

Everyone loves potato chips. They are sold the world over, with people in the United States of America and France the biggest fans. The number of potato chips consumed is mind-boggling. In the United States, Lay's alone sells 372 million bags each year. It's the go-to snack food of choice. Who made the first potato chip? That's a story worth chewing on.

While the birth of the potato chip is the stuff of legend, most origin stories revolve around one man by the name of George "Crum" Speck, a cook in Saratoga Springs, New York, who can also lay claim to being the United States' first celebrity chef.

George Speck was born in New York's Saratoga County in 1824, the son of Abraham Speck, a Black (some say mulatto) father, a jockey who migrated to the area from Kentucky, and Diana Tull, a Native American mother from the local St. Regis Mohawk tribe. Speck later would claim also to have German and Spanish bloodlines. As a teenager, Speck worked as a hunter and guide in the Adirondack Mountains just north of Saratoga. Slavery had been abolished in New York in 1827 and upstate New York was a link in the Underground Railroad for runaway slaves heading for Canada. Gerrit Smith, a wealthy white philanthropist, encouraged Black families to settle in the Adirondacks. John Brown, the abolitionist who was later executed for his raid on Harper's Ferry in Virginia on the eve of the Civil War, was living on a farm in North Elba in 1849 in support of Smith's efforts.

While the region was gripped in abolitionist fever, Speck found another passion. Speck earned a reputation in the culinary arts while in the Adirondacks, one that got him hired as a cook in the early 1850s at Moon's Lake House back in Saratoga Springs. In those days, Saratoga was well on its way to becoming the largest luxury resort town in nineteenth-century America, built on the European "spa" model to take advantage of the local mineral springs. Wealthy tycoons like railroad magnate Cornelius Vanderbilt, aka "The Commodore," built luxurious, mansion-sized "summer camps" in the area. Moon's Lake House was an elegant dining establishment built on a bluff overlooking a lake that quickly became the preferred destination for a daily parade of carriages when it was opened by Cary Moon in 1853.

Speck's specialty as a cook was the preparation of wild game like venison and duck, skills learned in the Adirondacks and sharpened by an early association with another well-known St. Regis Mohawk cook and guide named Pete Francis who worked at nearby Ballston Spa. But a house specialty on the Moon's Lake House menu was thick-cut French fried potatoes, an item popularized in the United States by Thomas Jefferson after his sojourn as the French ambassador many years before. The area had deep links to the American Revolution with the critical Battle of Saratoga fought in 1777, an association that bolstered the area's allure.

As the story goes, a grumpy patron (some say it was Cornelius Vanderbilt, a notion that may have its origins in the 1970s with an ad created for the Potato Chip/Snack Food Association) complained that his French fries were too thick and soggy and sent them back to the kitchen. A second serving, sliced thinner, was returned with the same complaint. Speck could be a bit surly himself and didn't take well to criticism of his cooking. In a pique worthy of any chef, Speck sliced the French fries paper thin and salted them with the intent of making them less tasty. Speck then deep fried the slices to make them crispy and impossible to cut with a knife and fork. To his surprise, the grumpy patron loved them and ordered a second serving. Soon other diners requested the "Saratoga Chips" and their popularity soared. Moon's Lake House had a new specialty. High-society ladies were soon seen strolling around town or in the paddock area of the local racetrack munching on Saratoga chips out of a paper sack as if they were candy or peanuts. "A gathered crowd was likely to create a sound like a scuffling through dried autumn leaves," wrote one observer.

And then there is the alternative origin version. Kate Wicks, Speck's sister and a line cook at Moon's Lake House, told a simpler tale that appeared in her 1917 obituary. Wicks explained that she had sliced off a sliver of potato and it inadvertently fell into a hot frying pan. Speck tasted the result and decided to serve them to diners. Researchers also note two American cookbooks, *The Cook's Oracle* (1817) and *The Cook's Own Book* (1832), have recipes for fried potato "shavings."

There is no doubt that Speck knew a good thing when he saw it. In 1860, Speck opened his own restaurant in Saratoga Springs and brought along a rich customer base that included Cornelius Vanderbilt, financier Jay Gould, and Judge Henry Hilton (who would achieve notoriety in 1877 by refusing to admit Jewish financier Joseph Seligman to his swanky Grand Union Hotel in Saratoga Springs).

Speck called his new restaurant "Crum's" with a focus on fish and game. Some say Speck named it after his father's nickname as a jockey. Other researchers note that Speck's paternal grandfather was named John Crum, who served in the Revolutionary War and fought in the Battle of Saratoga. It may be that the family used the names "Crum" and "Speck" interchangeably. A more colorful version stars Vanderbilt: he had trouble remembering Speck's name and mistakenly called him "Crum." Speck reportedly adopted Crum as a nickname, reasoning that a crumb was bigger than a speck. "Crum" clearly had a talent for self-promotion and keeping his name associated with a wealthy clientele. But in the parlance of the local horse racing track, Speck "played no favorites." Crum's had a "no reservations" policy: well-heeled customers standing in line for a high-priced meal was good advertising. Vanderbilt, so it was said, once stood in line for ninety minutes before being seated. Prices were on par with the best New York City restaurants of the day like Delmonico's but so was the food and service. Speck was an early farm-to-table disciple with much of the food coming from one of the farms he owned. Speck would be celebrated for his brook trout, lake bass, woodcock, and partridge creations. Meanwhile, potato chips appeared as a gourmet delicacy in tony hotels like the Cadillac in Detroit and aboard luxury cruise liners like the RMS *Berengaria*, served alongside the roast pheasant or accompanying the chicken salad.

Speck put his chips into baskets, placing them on all tables, and marketed a take-out box of Original Saratoga Chips. A more upscale side-dish preparation for expensive private parties came in special sterling silver "Saratoga Chip Servers" complete with grease-draining holes. Crum's became nationally known, and Speck became a celebrity chef known as George Crum. In 1889, a *New York Herald* writer dubbed Speck/Crum "the best cook in the country." Speck closed his restaurant in 1890 and passed away in 1914. (Moon's Lake House existed until the 1980s in various incarnations despite being burned down four times.)

Speck never patented his invention of the potato chip, likely because non-whites were discouraged from holding patents in the nineteenth century. Others took the potato chip and ran with it. In 1895, William Tappendum began making potato chips in a makeshift factory behind his house for direct sale to grocery stores, becoming the first of many to do so, where they were sold in barrels or out of glass display cases. In 1926, California entrepreneur Laura Scudder created the "bag of chips" concept by packaging them into wax paper bags. Herman Lay founded Lay's in 1932, selling his initial batch out of the trunk of his car, and it became the first nationally successful brand of potato chips. (Lay developed the "ruffled" chip that tended to be sturdier and less prone to breakage.) In 1954, Joe "Spud" Murphy's Irish Tayto company created the first seasoned potato chip (Cheese & Onion). In 1963, the advertising agency Young & Rubicam created the memorable "Betcha can't eat just one" slogan for Lay's. They were so right. Today, potato chip sales are calculated in the billions of dollars. The potato chip, once a dish associated with luxury dining, had become a snack food eaten by everyone.

4

Going Up

The Elevator Makes New York Skyscrapers Possible (1854)

Figure 4. Equitable Life Insurance Building showing elevator banks on left. *Source*: Grand Arcade, Equitable Life Assurance Building, New York. Library of Congress, Prints and Photographs Division, Detroit Publishing Co. no. 013293.

Not many inventors have to put their life on the line to prove that their invention works but that's precisely the precarious position Elisha Otis put himself in to demonstrate that his elevator was safe. Otis stood suspended on a platform high above a crowd egged on by P. T. Barnum of circus fame, with an axe in his hand, ready to cut the cable. The crowd believed Otis would plunge to his doom. Otis cut the cable and the platform remained where it was. "All safe!" cried Otis. This one act of courage and showmanship paved the way for a New York cityscape of skyscrapers that would become synonymous with urban centers around the world.

Elisha Graves Otis's journey began on August 3, 1811, when he was born to parents Stephen and Phoebe in Halifax, Vermont. At nineteen, Otis moved to the Albany area and worked as a wagon driver. While his talent as an inventor was recognized early on, Otis experienced a run of bad luck. A bout with pneumonia nearly killed him. Otis married but his wife died, leaving him with two small sons. An early patent for a device that increased doll production earned him enough money to start his own business, one that was soon scuppered when the city of Albany diverted the water he needed for a power supply.

Looking for a fresh start and now remarried, the forty-year-old Otis moved south to become the manager of an abandoned sawmill in Yonkers, New York, that was to be converted into a factory for the production of bedsteads. Otis's main problem was how to move heavy debris between floors. Hoisting platforms dated to the construction of the pyramids but these broke so often that they were unsafe for people to use and dangerous to operate even for freight. No one wanted to be under one if the cable broke.

Otis's solution was ingenious. Otis attached saw-toothed ratchet bars to each of the four guide rails and placed a wagon spring atop the platform. When the lifting cable was attached to the upper bar of the spring, the pull from the heavy platform was taut enough to keep the platform from touching the guide rails. However, if the cable snapped, the tension on the spring would be released and the ratchet bars would lock the platform in place to prevent it from falling.

The plan for a bedstead factory fell apart so Otis opened his own factory in 1853 on the site for the construction of his safety elevator. The trick was getting noticed, which is why Otis found himself above the crowd at the magnificent iron and glass Crystal Palace Exhibition

Hall—where Bryant Park on Forty-Second Street is now—hosting the 1853–54 World's Fair. The event was not overly successful in its first year, so organizers turned to the famous showman P. T. Barnum for help. Barnum paid Otis $100 to erect and demonstrate his invention. When Otis, sporting a heavy beard that would make any lumberjack proud, dramatically cut the supporting cable with his axe, the platform moved only a few inches before coming to a stop, shocking the crowd. Otis's safety mechanism prevented the platform from crashing to the ground.

This bit of elevator derring-do was no guarantee of success, but the arrival of a safe passenger elevator was expected by key figures in the architectural trade. Four years earlier, the architect Peter Cooper designed New York's Union Foundation Building with a cylindrical shaft, confident that passenger elevators were on the horizon.

The first passenger elevator, however, was installed in 1857 in a new five-story building at the corner of Broome Street and Broadway in Manhattan. The E. V. Haughwout Building sported two cast-iron facades and was home to a high-end emporium selling imported glass and silverware as well as chandeliers and its own painted china—Mary Todd Lincoln would later source White House china from here.

The now landmarked Haughwout Building was notable in that it used cast-iron facades as a structural frame, foreshadowing the steel-framed skyscrapers to come in the next century. Otis installed a passenger elevator, which was powered by a steam engine in the basement. The elevator had a speed of about forty feet per minute in contrast to modern elevators that cover that distance in a second. A five-story building didn't really need an elevator, but Haughwout knew people would come to experience the novelty and buy merchandise while they were at it. The Haughwout elevator was the world's first, but it has since been removed.

The passenger elevator proved central to the construction of the block-sized Equitable Life Building in 1870 in lower Manhattan. Originally designed with two sub-basements and seven floors, two were later added. The elevator turned real estate on its head. Prestige law floors had never rented offices above the second floor to spare workers the potential exhaustion of climbing the stairs. Now the upper floors became desirable as higher floors had better sunlight, ventilation, and views. The luxury penthouse would appear in the 1920s, turning the most undesirable space under the roof into a terraced heaven.

Elisha Otis wouldn't live to see it however, dying from diphtheria in 1861 at the age of forty-nine after narrowly avoiding bankruptcy just a few years before. Otis's sons took over the business with Norton traveling the country to promote Otis Elevator while Charles ran the factory in Yonkers. The complete elevator would be built in Yonkers, then disassembled for shipment to the customer where a local installer would re-assemble it for installation into a shaft. Otis eventually built a special cylinder model for Peter Cooper. Steam engines gave way to elevators powered by faster hydraulic systems and safety continued to improve. One big safety innovation came from outside the company: in 1887, Black inventor Alexander Miles patented a widely-adopted method for automatic elevator door operation so they couldn't be left open accidentally. Elevators were advertised as being safer than the stairs.

Elevators got a lift after the Civil War with many becoming sumptuous ascending rooms for hotels and fine stores. The Congress Hotel in Saratoga, for example, in 1870 installed an elevator with skylights, gaslit chandeliers, sofas on three sides made with fancy wood trims and other gilded touches, with operators decked out in uniform and gloves.

By 1884, Otis Elevator had installed 1,250 elevators in New York City alone. When the ten-story Home Insurance Building, commonly considered the first skyscraper due to its steel frame, went up in 1885 in Chicago, it signaled that elevator shafts had become a core element of architectural design. But Otis Elevator's reputation was sealed in 1889 when it alone, despite French opposition to the use of American technology, figured out how to run elevators up the curved legs of the new Eiffel Tower in Paris. Otis Elevator switched to elevators powered by electricity and operated by push buttons in the early 1890s. Competitors emerged but Otis Elevator swallowed thirteen of them and formed partnerships with many others. In 1898, Otis Elevator went public, one of the earliest companies to do so other than railroads, with a market capitalization of $11 million. At the same time, Otis Elevator acquired the rights to a new invention that would ultimately be called escalators, installing the first one at a subway station at Sixth Avenue and Twenty-Third Street in 1899.

Norton retired from Otis Elevator in 1890 and entered politics, serving as a mayor of Yonkers and as a Republican congressman from New York's Nineteenth Congressional District. Charles stayed on at Otis Elevator until the death of his brother in 1905 when a trusted associate, William Baldwin, took the helm through 1930. Demand soared. Otis Elevator built elevators for the London Underground and at the Kremlin for the Czar of Russia. The company operated seven factories in the United States and one in Canada. If you wanted to go up, Otis Elevator got you there. The Empire State Building was a high point when it opened in 1931, with seventy-three Otis elevators serving 102 floors and traveling at the unprecedented speed of 1,200 feet per minute. The Empire State Building would also be the scene of the most bizarre accident involving Otis elevators. A B-25 bomber crashed into the building in 1945, severing the cable of one lift that promptly plunged from the thirty-eighth floor to the basement (operator Betty Lou Oliver survived the fall). Another high point was the installation of 255 Otis elevators and 71 escalators in the World Trade Center in 1968. But even as elevators went into high-rise buildings, they became increasingly popular in low-rise structures of just a few floors.

Otis Elevator would soon become part of an international conglomerate called United Technologies Corp. but was subsequently spun off into an independent publicly-traded company. Elevator operation is increasingly automated to the point where they can now be compared to a self-driving car. The illusion of control is the elevator's dirty little secret: the close door button is usually inoperable even though it is likely the most pushed button in elevator history. Doors are set to remain open for a specific amount of time to ensure access by wheelchairs and people with mobility issues.

Certainly, the prospect for elevators is looking up. Elevators that move sideways—moving sidewalks—are a growth business. One extreme proposal is for a space elevator that would send passengers 22,000 miles up to a fixed geostationary spot orbiting the globe. That idea may seem as fantastic as the one Elisha Otis demonstrated to an awed crowd in 1854.

5

The Submarine Paid For by Irish Rebels (1881)

Figure 5. The USS *Plunger*, c. 1905. *Source*: Library of Congress, Prints and Photographs Division, LC-USZ62-89935.

When the modern-day submarine first set sail, it was a revolution in modern warfare. Fittingly, the first of these submarines was designed to foment an actual revolution.

The desire for a vessel that could travel underwater is probably as old as the first ship that could travel on the surface of the sea. According to legend, Alexander the Great reportedly used a glass diving bell to conduct reconnaissance during the Peloponnesian Wars of 332 BC. In the ensuing centuries, inventors never lost interest in the idea, but most efforts proved impractical. During the American Revolution, an egg-shaped, single-person, hand-powered submersible called the *Turtle* took to the waters in New York Harbor in a failed bid to sink an English ship. In the American Civil War, the *Hunley* became the first submarine to sink an enemy warship. Military interest quickened after the war with the development of the Whitehead torpedo propelled by compressed air and equipped with an explosive warhead. Public interest in submarines was stoked by the fictional *Nautilus* captained by Nemo in Jules Verne's 1870 novel *Twenty Thousand Leagues Under the Sea*.

Among those excited by Verne's novel was John Holland, a slight Irishman with poor eyesight and a penchant for bowler hats, newly arrived in the United States. Holland, whose father had been a member of the Coast Guards in Ireland, had dabbled with submarine designs as early as 1859 while still in Ireland. Holland's poor eyesight prevented him from following his father to the sea, but ship design became a passion. Holland had joined the Christian Brothers religious order and taught math at a local school, but due to ill health he was released from his vows. Holland's submarine designs would come to fruition in New York; however, in 1873, Holland was in Boston, joining his family who had immigrated a few years before, and working for an engineering firm. While in Boston, Holland slipped and fell on thin ice, suffering a broken leg and a concussion. Holland, who later said the accident was the luckiest thing that ever happened to him, used the recuperation time to fine tune his submarine design. In 1875, Holland offered his designs to the United States Navy. The response was very discouraging. Holland's submarine was the "fantastic scheme of a civilian landsman," they said in rejection.

Alongside thousands of other Irishmen who resented England's colonial rule in Ireland, Holland was strongly anti-British and believed England's mastery of the seas was the strongest obstacle to Irish independence. Holland's younger brother Michael was associated with the well-funded Fenian Brotherhood movement in the United States that opposed British rule. With an introduction from Michael, Holland convinced the rebel Fenian leaders his submarine would undermine British naval superiority. The Fenians agreed to use their "skirmishing fund" to finance Holland's construction of a submarine. The idea was to construct a three-man submarine that would be carried aboard a harmless-looking merchant ship. The submarine would slip out through an underwater door, carry out its attack, and return to the mother ship like a sea-going Trojan horse. The Fenians allocated six thousand dollars in start-up funds. The Fenians were no strangers to daring schemes, having mounted three clandestine invasions of Canada with an army of Irish Civil War veterans between 1866 and 1871 with the aim of swapping captured Canadian territory back to England in exchange for Irish independence, a scheme that proved unsuccessful due to Canadian opposition by force of arms.

Holland, who by this time was teaching school in New Jersey, built a fourteen-foot, one-man submarine powered by a four horsepower Brayton engine as a proof-of-concept and demonstrated its viability with a dive to twelve feet in the Passaic River in 1877. The excited Fenians advanced further funds and Holland went to work on the submarine project at the DeLamater Iron Works on West Fourteenth Street in New York City. Delamater was well known in naval circles, having produced the famous ironclad *Monitor* during the Civil War. Holland eventually delivered a thirty-one-foot submarine in the summer of 1881 that held a crew of three.

Crucially, the submarine used a small Brayton gas engine for surface sailing and a battery-powered electric motor for underwater propulsion, a basic configuration that submarines would utilize until the development of nuclear-powered submarines decades later. One man sat above the engine, using two joystick levers to control the rudder and diving planes. An engineer operated valves and monitored gauges while underway. The third crew member was in charge of the pneumatic gun, an eleven-foot-long tube that fired a six-foot-long torpedo with a burst of compressed air. During trials, the submarine reached depths of forty-five feet and had a surface cruising speed of nine knots. Blakley Hall, a reporter for the *New York Sun,* dubbed it the *Fenian Ram.*

The *Fenian Ram* was a success. Holland surmised the *Fenian Ram* could stay submerged for three days and fire a torpedo fifty yards underwater. Two features would be key to future designs: a water ballast to submerge the sub and horizontal rudders for diving. Holland continued to improve on his original plans and a third, more seaworthy, nineteen-ton boat was soon in development.

The Fenians, though, were worried. Construction costs had risen to sixty thousand dollars, and they were no closer to sinking a ship of the British Navy. The Fenians accused Holland of misusing funds and Holland claimed the Brayton engines had cost more than anticipated. Lawsuits were filed but the Fenians worried that these legal problems might lead to the seizure of the submarine. The Fenians forged a pass with Holland's name on it and clandestinely moved the *Fenian Ram* by tug to Connecticut. The relationship between Holland and the Fenians dissolved into a bitter divorce, with the Fenians claiming physical ownership of the submarine, much to Holland's annoyance. The *Fenian Ram* languished in storage, only to be used years later in a fundraising campaign for the families of the 1916 Uprising in Ireland. Today, the *Fenian Ram* resides at a museum in Patterson, New Jersey.

Holland, meanwhile, continued to work on submarines privately. Holland formed a partnership with Edmund Zalinski, an army officer with a reputation as a developer of military technology. The venture was named the Nautilus Submarine Boat Company, a nod perhaps to Jules Verne. Funding was hard to come by. The *Zalinski* boat was built on a shoestring budget in the Bay Ridge section of Brooklyn. The hull of the fifty-foot boat was made largely of wood held in place by iron hoops. Zalinski added a pneumatic gun of his own design that fired projectiles using air pressure while also rejecting most of Holland's suggestions on improving the overall submarine design. The initial 1885 launch didn't go well: a section of the way (or framework) holding the *Zalinski* in place prior to launch collapsed under the weight of the submarine and damaged the hull. Repairs were made but investor interest never developed, and the partnership dissolved.

Holland was forced to take a job as an engineer to make ends meet. But in Washington, DC, naval interest in submarines slowly grew. The United States Navy sponsored a design competition, which Holland won. With the backing of a wealthy lawyer named E. B. Frost, Holland set up the John Holland Torpedo Boat Company in 1896. The navy, however, still had its doubts about Holland, calling him a "gifted amateur" at best. The navy insisted on some radical changes to Holland's initial design for what became known as the *Plunger*. Holland predicted the alterations would not work and he was proved right. The initial project was abandoned but Holland came to a critical judgement: steam engines in a submarine would never work over the long haul. Holland also needed a success. He was newly married to Margaret Foley, the daughter of an Irish immigrant, with whom he would have five children.

The navy had learned its over-engineering lesson. *Holland VI* proved to be the inventor's most successful effort thus far. Constructed across the Hudson River at the Crescent Shipyard in Elizabeth, New Jersey, the fifty-three-foot vessel was fitted with a single torpedo tube, came to the surface quickly, and took only five seconds to disappear beneath the waves. The submarine made its first dive on St. Patrick's Day in New York Harbor. Holland tried to attract as little attention as possible, but the New York press was already agog over "the terror, the wonder, and the monster." Rumor had it that both Spain and France were interested in acquiring the *Holland VI*. In the background, tensions with Spain were high over the sinking of the *Maine* in Havana Harbor as the United States headed toward the short-lived Spanish-American War.

Holland shifted operations to Brooklyn, first to the Erie Basin and then to the Atlantic Yacht Club. More trials were conducted in the Narrows below Bay Ridge and in the Lower Bay; voyages often conducted with influential naval personages, both domestic and foreign, on board. It wasn't all celebrity cruising though. Marked improvements to steering and firing mechanisms were made as a result.

The busy waters of the city proved too chaotic for extended trials, so operations moved again to the Goldsmith and Tuthill Shipyard in New Suffolk on Long Island. A three-mile course in Little Peconic Bay was completed and the occasional celebrity like Clara Barton, founder of the American Red Cross, went along for the submerged ride. At this point, Holland could taste success as surely as he could the salt on an ocean breeze.

The final test was before the Naval Review Board at the Washington Navy Yard on the Potomac River. During subsequent maneuvers off Norfolk, the submarine and its six-man crew proved undetectable by warships searching for it. Admiral Dewey, one of the witnesses and himself a naval hero for actions in the Philippines during the Spanish-American War, said: "If they had that sort of thing at Manila, I never would have held it with the squadron I had. The moral effect is immense. It is wholly superior to mines or torpedoes."

Secretary of the navy and future president Theodore Roosevelt lent his support, and the US government bought its first submarine on April 12, 1900, for $150,000, a bargain as it had cost twice that amount to build. The *Holland VI* became the USS *Holland*. The inventor had persevered despite decades of discouragement.

The United States Navy ordered six more submarines, but it was not to be Holland's only customer. With a patent secured in 1902 and the company renamed the Electric Boat Company, Holland built two submarines for Japan that were used in its 1904–05 war with Russia and they contributed to the Japanese victory. Holland was decorated by the Japanese emperor for his efforts. Austria built submarines for many countries under license. Holland died at seventy-three in 1914, a few days before the outbreak of World War I. A few weeks later, a German submarine called the U-9 with a crew of twenty-six would sink three British cruisers with the loss of fourteen hundred men. The Electric Boat Company still exists today under the name General Dynamics Electric Boat in Groton, Connecticut. Among the numerous submarines built by the company was the first nuclear-powered submarine, the USS *Nautilus*, in 1954.

Holland was buried in New Jersey about a mile from where he launched his first small submarine model for the Fenians. Ironically, Holland's tangle with the Fenians years before clearly left him with bad feelings toward them even after decades had passed. Holland sold his submarine designs to the British Navy and the Fenians never took to the sea.

6

The First Roller Coaster (1884)

Figure 6. Scenic Railway entrance, Luna Park, 1906. *Source: Luna Park*, 1906, Gelatin silver print, WEML_0011; Eugene Wemlinger photograph collection, Brooklyn Public Library, Center for Brooklyn History.

A modern-day roller coaster drops riders from heights well above four hundred feet at speeds over one hundred miles an hour in a screaming thrilling ride that last seconds. That's a long cry from the first one that debuted in Coney Island in 1884. Riders were leisurely swept along at six miles an hour, from a fifty-foot height across a six-hundred-foot-long track, for a ride that lasted several minutes. Then, like now, they loved it. The ride transformed a Brooklyn resort into the world's most famous amusement park.

The man who brought the roller coaster to Coney Island wasn't a New York native. LaMarcus Adna Thompson was born in Jersey, Ohio, in 1848. As a young boy, Thompson built mechanical toys and exhibited a knack for carpentry, building a large barn for his father at the age of seventeen. Thompson entered Hillsdale College in Michigan but only stayed for one semester as he couldn't afford the tuition. Thompson moved to Indiana where he developed a career as a successful businessman, inventing a knitting machine for the manufacture of seamless stockings for women that netted him a fortune. The task of putting hosiery on female legs proved too stressful for Thompson, however. While still in his early thirties but on the verge of a nervous breakdown, Thompson sold his Eagle Knitting Company and headed to Arizona for six months to rest and recuperate.

Thompson was a devout Christian and believed Americans were increasingly being lured toward wicked and hedonistic entertainments offered by saloons and brothels. The country was in the midst of a spiritual crisis, he believed. Thompson decided to devote his energies toward the creation of more wholesome entertainment.

Inspiration came from the Mauch Chunk Gravity Railway, a nine-mile downhill railway in Pennsylvania that hauled coal out of a mountain mine to distant loading docks. The acceleration came from gravity with as many as fourteen cars loaded with coal speeding downhill, guided by a courageous "runner" operating a brake lever. Mules, and later steam engines on a back track, would haul the train back up the mountain. Most of the coal was hauled in morning runs so the adventurous, Thompson among them, paid fifty cents for an afternoon ride on "gravity road."

Thompson targeted New York's Coney Island as a place desperately in need of wholesome entertainment. By the 1880s, Coney Island had been transformed from a sleepy farming community into a massive sea-side resort. Railroad companies and other major investors spent millions developing hotels, music venues, restaurants, and private bathing houses where visitors could rent a bathing suit. The idea of swimming in the ocean as a recreational activity was still a new one. Jules Verne's *Twenty Thousand Leagues Under the Sea* was published in 1870 and was still fresh in people's minds. More than a few thought sea monsters called the ocean home, not to mention futuristic submarines disguised as such.

With the help of railroad companies that wanted to boost weekend ridership, Coney Island became a tony destination with establishments like the Manhattan Beach Hotel, the Oriental Hotel, and the Brighton Beach Resort catering to the wealthy with a racetrack, top music acts, and elaborate fireworks displays. Among the well-known clientele were circus master P. T. Barnum, poet Walt Whitman, and writer Herman Melville who reportedly worked on his *Moby Dick* masterpiece while staying at the Coney Island House.

Inevitably perhaps, Coney Island also attracted a seedier clientele. The Manhattan Beach Hotel allowed co-ed bathing, shocking for its time, but barred Jews and other minorities. At the west end of the island, a section known as the Gut was an unregulated ten-block area hosting betting parlors, opium dens, dance halls, boxing rings, and bordellos. Holes were drilled into old bathhouses, transforming them from changing rooms to one-dollar-a-look peep shows. The "la-de-da boys" performed at illegal drag cabarets. Bribed police looked the other way. The reputation of Coney Island was such that when Charles Feltman debuted the hot dog there in 1867, it had to be pitched as "Coney Island Chicken" to assure customers they were not eating canine on a bun. Preachers and journalists railed against Coney Island as a four-mile-long citadel of crime, intoxication, and debauchery. Coney Island helped give New York City a reputation for corruption.

It's unclear whether Thompson arrived in Coney Island at the invitation of a local railroad magnate or was drawn to Coney Island because he saw himself as its spiritual savior. But Thompson stepped into this environment willingly, opening the Gravity Pleasure Switchback Railway on June 16, 1884. The Switchback Railway was a six-hundred-foot-long ride along mild, undulating hill tracks with a beach view that sped along at six miles an hour. Passengers rode between two fifty-foot towers, getting off midway so the carriage could be turned around for the return trip, all for five cents. On an island full of thrill seekers, a ride that allowed passengers of the opposite sex to hip rub each other was the Love Boat of its day. Thompson's ride was a hit. Thompson recouped his $1,600 investment in just three weeks.

Thompson was delighted. "Many of the evils of society, much of the vice and crime we deplore come from the degrading nature of amusements . . . to substitute something better, something clean and wholesome, and persuade men to choose it, is a worthy endeavor," said Thompson. "We can offer sunshine that glows bright in the afterthought and scatters the darkness of the tenement for the price of a nickel or a dime."

Thompson's roller coaster set the stage for Coney Island to become the mother of all amusement parks. Within months, Thompson had competitors, many of whom introduced refinements like a mechanical hoist to lift cars and a circular loop track. Side-facing benches were replaced by forward-looking seats. The basic elements of the modern roller coaster were in place.

While modern roller coasters emphasize speed and steep drops, Thompson chose a more scenic route to success in an era when travel was difficult and costly. While initially built to give roller coaster riders a view of the surrounding landscape, Thompson hit on the idea of incorporating dioramas and painted scenic landscapes so riders could feel like they were on a rolling tour through the Swiss Alps and other foreign locales. The elaborate artificial scenery would be illuminated by the lights triggered by the approaching cars, foreshadowing the development of modern theme park rides operated by Disneyland, for example. The first scenic roller coaster, a partnership with designer James A. Griffiths, debuted on the Atlantic City boardwalk in New Jersey. The L. A. Thompson Scenic Railway Company would soon operate six scenic railways on Coney Island alone, with the flagship ride being the Oriental Scenic Railway. Thompson also developed a scenic coaster that traveled through tunnels illuminated only by the headlights on the lead car—the first Tunnel of Love. Thompson would go on to operate fifty scenic coasters in the United States and Europe and become a high-profile celebrity whose whereabouts were followed by newspapers. Every scenic railway would be more elaborate than the previous one. Among the most notable is one in Venice, California, that opened in 1910, which tracked through replicas of temples illuminated by artificial lights. If you wanted to be swept away into another world, Thompson's scenic roller coaster brought you there. Thompson would stay firmly rooted on Coney Island with the company's main office and workshop on Surf Avenue and West Eighth Street.

Thompson, the holder of thirty patents, also would train a small army of roller-coaster designers and safety improvements made the rides even more popular. Roller coasters would become the central fixture of legendary amusement parks with names like Luna Park, Dreamland, and Steeplechase Park. John Miller, Thompson's chief engineer, would lay claim to over one hundred patents, the most important being a safety mechanism that prevented cars from rolling backwards down the lift hill if the pull chain broke and underfriction wheels that kept coaster cars locked onto their tracks, a development that fostered higher speeds and allowed cars to turn upside down or bank suddenly. These safety improvements and others eventually saw scenic coasters eclipsed by more thrilling rides. By the 1920s, considered the golden age of roller coasters, fifteen hundred were in operation across the United States with an equal number overseas. Rivals to Coney Island, like Chicago's Riverview Park, promoted roller coasters like the Fireball with an alleged speed of one hundred miles an hour, circumventing seventy-two-foot height limits in city ordinances by building the first drop into a man-made ditch. Coney Island was by then reachable by subway and summertime visits were in excess of one million people per day. Of the roller coasters of that era, the most memorable one still standing is Coney Island's Cyclone, built by the Harry C. Baker Company in 1927 and based on a design by Vernon Keenan. The sinuous, sixty-mile-an-hour Cyclone is constructed of wood and has since achieved iconic landmark status as the "Big Momma" of roller coasters with an eighty-five-foot plunge and twenty-seven elevation changes. "A ride on Cyclone is a greater thrill than flying an airplane at top speed," said Charles Lindbergh, the first aviator to fly solo across the Atlantic Ocean in 1927. The Cyclone occupies the same spot as Thompson's original roller coaster.

Roller coasters suffered through a decline during the Great Depression and World War II when many were forced to close. The Monticello Amusement Park in New York's Sullivan County, with a roller coaster called Pippin—"the second largest in the state"—opened in 1923 but fell victim to a devastating fire of suspicious origin in 1932 and was never rebuilt. In 1955, the opening of Disneyland in southern California signaled a second golden age for roller coasters. Technological innovations like tubular steel construction now allow roller coasters to soar to dizzying heights of over four hundred feet and have made them more popular than ever.

Thompson died at his home in Glen Cove, Long Island, on May 8, 1919, at the age of seventy-one. Historians note that gravity-powered rides date back to at least the seventeenth century when ice slides were popular in Russia. But there is no doubt that Thompson is the father of the American roller coaster for his role in popularizing an entertainment that continues to be enjoyed the world over.

7

Photography's Kodak Moment (1887)

Figure 7. "Kodak Girl": Woman holding early Kodak camera. *Source*: No. 28—The Kodak Girl. Library of Congress, Prints and Photographs Division.

Kodak. The word meant nothing when George Eastman coined it in 1887 but it quickly became synonymous with photography and has remained so for a century. Eastman put photography into the hands of everyone at an affordable price and the promise of easy operation. "You press the button, we do the rest," claimed the first Kodak camera ad. That sentence set the standard for photography ever since. In what seemed like an instant, Eastman created personal photography, a startling achievement for a man who was a high-school dropout. More startling is that Eastman ended his life in suicide.

Just how cumbersome photography was before Kodak was detailed by Eastman. In 1874, Eastman made plans to go to Santo Domingo in the Dominican Republic amidst speculation that the United States would soon annex the Caribbean island. A friend suggested Eastman record his trip with a photography outfit. The "outfit" included a heavy camera and tripod, chemicals, glass tanks, plate holders, and a jug of water. Eastman also would need a tent to spread the photographic emulsion on glass plates before exposing them and then develop the plates before they dried out. It was a "pack-horse load," Eastman recorded.

Eastman never made the trip, but he became obsessed with making photography easier. The photographic wet plates in use at that time held little promise in that regard. But Eastman soon learned that British photographers were experimenting with dry emulsions for photographic plates that retained their sensitivity and could be used at leisure. Eastman began his own experiments at night in his mother's kitchen, as he was unable to ditch his day job as a clerk at a bank. Some nights, an exhausted twenty-four-year-old Eastman would just sleep on a blanket by the kitchen stove.

Eastman's obsessive personality traits can be traced to an unfortunate childhood. Eastman was born on July 12, 1854, in Waterville, New York, not far from the city of Rochester that would become Kodak's base of operations. Money was tight. His father George Washington Eastman worked two jobs to make ends meet but died when George was eight, leaving the family impoverished. One of his two sisters died from polio in 1870. At fourteen, Eastman dropped out of high school to help support his family.

Those years left a mark on his personality. Eastman rarely spoke of his childhood but in later years a close friend, Reverend George Norton, recalled Eastman saying "I never smiled until after I was forty. I may have grinned, but I never smiled."

Eastman's mother Maria, a difficult-to-please Calvinist, opened their home to boarders for much-needed income. Perhaps the only stroke of good fortune for Eastman was that the boarders were the Strong family. Eastman and Henry Strong struck up an unlikely friendship. Strong was gregarious while Eastman was taciturn. Strong, a navy paymaster during the Civil War, would become a key early investor and later the president of Kodak. Together, they were a formidable pair.

Eastman, meanwhile, vowed to make up for all the deprivations his mother had endured. It took Eastman three years to hit upon a dry emulsion formula that worked. "Finally, I came across a coating of gelatin and silver bromide that had all the necessary photographic qualities," Eastman told a photography journal. "At first, I wanted to make photography simpler merely for my own convenience but soon I thought of the possibilities for commercial production."

Eastman realized that there was a need for a machine that could cheaply and uniformly turn out the dry plates. Eastman developed one and patented it. Eastman also designed and patented ruby-red light bulbs for a darkroom, which were custom-made at Corning Glass Works, also in upstate New York, a company that would soon become famous as the mass producer of Thomas Edison's light bulbs.

In December 1880, Eastman and Strong set up the Eastman Dry Plate and Film Company. Eastman eventually quit his day job, bought land, and constructed a four-story building for operations. But even before the building was finished, Eastman was nearly ruined. Some plates began to produce badly fogged images or no images at all. As it turned out, a trip to England revealed that the chemical supplier had changed its gelatin source without informing them. Forty years later, Eastman would learn that the problem arose because the cows that the gelatin was made from had not grazed enough on sulfur-rich mustard.

Eastman continued to work on cameras, developing an award-winning roll-holder system for paper film that eliminated individual glass plates and reduced the weight of a camera from fifty pounds to two and three-quarter pounds. The problem was that there weren't enough photographers to create a large business. George Eastman, by now known as GE, hired the best university-trained minds he could find for technical research while he turned his thoughts toward marketing. In 1887, Eastman launched a reliable box-style film camera that sold for twenty-five dollars. The marketing stroke of genius was that the company would develop the one-hundred-picture roll of film for ten dollars and send the camera and developed photos back to the camera owner. No prior knowledge of photography was required. To use the camera, one armed the shutter by simply pulling on a string, pointing the camera at the subject, and then pressing the shutter release. The "You press the button, we do the rest" idea was born. The plan, said Eastman, was "to make the camera as convenient as the pencil." Kodak entered the lexicon, with the word being used as a noun, verb, and adjective. Fans became Kodakers.

Kodak was a hit. Eastman began to attract talent with stock options rather than high salaries, a novel idea at the time. The company changed its name from Eastman Dry Plate and Film Company to Eastman Kodak. Steady improvements to the camera were made every year but Eastman soon realized the money was in the sale of film.

A Kodak chemist named William Reichenbach developed a transparent flexible film that could be cut into strips and inserted into a camera. This film, used by Thomas Edison in his early experiments with the motion-picture camera, became the centerpiece of the Kodak empire, becoming a more important source of income than consumer film. Expansion abroad began with the opening of a new plant in England—GE took his beloved mother along for a three-month visit. By 1895, Kodak was making ninety percent of the film produced in the world. In subsequent years, Kodak would be a leader in X-ray films, aerial photography for military use, and synthetic chemicals.

The corporate emphasis on film did not mean the camera was forsaken. A big breakthrough was the debut of the famous Brownie camera in 1900 that sold for one dollar, with the film costing fifteen cents. Within the blink of a shutter, hundreds of thousands were sold.

Eastman's world was shaken in 1907 with the death of his mother after a long illness that confined her to a wheelchair. Eastman, a man normally in control of his emotions, was devastated. "When my mother died, I cried all day," Eastman admitted.

Eastman was a confirmed bachelor, but he did develop an intense and by all accounts platonic relationship with Josephine Dickman, trained singer and the widow of the man who managed Kodak's London office and who had died suddenly during Eastman's visit. Eastman's mother liked Josephine who would often look after her while Eastman traveled. With his mother's passing, their relationship deepened, and Josephine became a frequent travel companion. Eastman loved to photograph her in every setting imaginable. Eastman's biographer, Elizabeth Brayer, opined: "Considering his shyness, perhaps all this photographing was a way of making love without making love."

By 1913, Eastman was one of the richest men in America, but he was a man of contradictions. While Eastman was painfully shy and a man of "expressive silences," he often traveled in large groups of friends that were lavishly entertained. At times suspicious, Eastman would send spies into competitors' operations to see if they were using his patents. Detectives were hired to determine if salespeople were following instructions. Eastman shunned publicity and, ironically, the number of photographs of Eastman is relatively few considering his stature at the time. Eastman could walk down a street in Rochester in anonymity. More recognizable perhaps was the 42,000 square foot home Eastman built with thirty-seven rooms and thirteen bathrooms, with a staff of forty to take care of the interior and another fifteen for the extensive grounds. It is now the George Eastman Museum and a repository of photography and film.

Loyalty was amply rewarded. Kodak was a leader in employee benefits, offering health care and insurance, recreational areas, and pension benefits. A wage dividend to employees, begun in 1912, which awarded two percent of wages over the previous five years was turned into a Rochester retail holiday known as Kodak Bonus Day. Employees also could buy stock. Executives were habitually promoted from within the company. Rochester became a company town. The red and yellow logo was everywhere.

Likewise, Eastman's generosity was legendary, to the tune of millions of dollars to academic institutions like the Massachusetts Institute of

Technology, University of Rochester, and minority colleges like Hampton and Tuskegee. GE also established the Eastman School of Music in Rochester as well as remedial dental clinics for children in several countries, the latter being the most rewarding of his philanthropic endeavors, claimed Eastman.

"If a man has wealth, he has to make a choice because there is money heaping up," said Eastman. "He can keep it together in a bunch and then leave it to others to administer after he is dead. Or he can get into action and have fun while he is still alive. I prefer getting it into action."

By 1921, at the age of sixty-seven, Eastman dialed back his daily visits to the office and officially retired in 1925. Some quirkiness remained, however. Eastman became enamored of a plan to institute an allegedly more efficient twenty-eight-day, thirteen-month "Liberty" calendar year (the thirteenth month would have been called Vern and slipped in between June and July) and lent it substantial support. Yearly trips to Europe yielded a substantial art collection. And he remained an avid outdoorsman.

By the age of seventy-seven, Eastman suffered from a debilitating condition that hardened the cells in lower spinal cord and reduced his gait to a slow, painful shuffle. The memory of his mother's final years surely came to mind. On an icy March 14, 1932, after entertaining visitors and signing his will, Eastman took off his glasses, laid down on his bed, put a folded wet towel over his heart, and shot himself with a Luger automatic pistol. On a sheet of lined yellow paper on a bedside table a note read: "My work is done. Why wait?"

Eastman's last message made newspaper headlines around the world. Eastman was cremated and his ashes buried on the grounds of Eastman Business Park in Rochester.

Kodak as a company would thrive through the twentieth century with products like a variety of 35mm films such as Kodachrome that would capture countless "Kodak moments," Kodak introduced 16mm and 8mm home movies, the Instamatic camera, the photo CD, and more. In 1973, superstar singer Paul Simon sang "Mama, don't take my Kodachrome away!" By 1998, Kodak employed 145,000 people worldwide.

Kodak didn't falter in the eyes of consumers until the arrival of the digital age. Twenty-four-year-old, Brooklyn-born Kodak engineer Steve Sasson developed the first digital camera in 1975, an invention that turned out to be the engine of Kodak's destruction. Kodak executives killed it, fearing a digital camera would cannibalize the existing film business. "Why would anyone want to look at their pictures on a television set?" mused Kodak executives, as recounted by Sasson in a *New York Times* interview in 2015. Sasson's counter-argument was that the day would soon arrive when Kodak would be unable to sell film. In the interim, companies like Fuji and Polaroid made inroads in the 1980s with film that was easier to develop than Kodachrome

Sasson's patent expired in 2007 and competitors quickly moved in. That year, Apple introduced the iPhone with a built-in camera and changed photography forever. In 2012, Kodak filed for bankruptcy, a modern story of missed opportunities. The cash cow that was film couldn't be replaced. Today, Kodak operates in a more diminished capacity in specialty markets, including film production, and is enjoying a revival in film photography similar to that of vinyl records. But the company is no longer the monopolistic power of yore. A big revenue generator is the sprawling Eastman Business Park that now plays host to other companies. Rochester's fortunes declined as well when the paternalistic system that was the city's foundation disappeared.

Sasson, incidentally, was awarded the National Medal of Technology and Innovation by President Obama in 2009. What Eastman would make of him is worth pondering.

8

The Machinery of Democracy (1892)

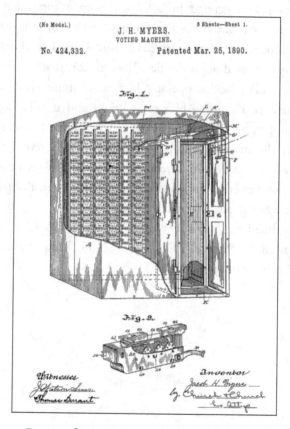

Figure 8. Patent Drawing for J. H. Myers's voting machine. *Source*: United States Patent and Trademark Office, US Patent 415,549, issued November 19, 1889.

The election was close, and the losers didn't accept the results. Witnesses to the 2020 presidential election between incumbent Donald Trump on the Republican side and former vice president and Democrat Joe Biden will likely not forget the scenes of Trump supporters seemingly ready to storm polling stations as they disputed an outcome that resulted in a Biden win. Trump drafted executive orders to seize control of voting machines in key states and inquired as to their feasibility with the Pentagon, the Justice Department, and the Department of Homeland Security, all of which declined to act. Federal intrusion into elections run by states was uncharted waters. Trump also encouraged local law enforcement to seize voting machines in Michigan and Pennsylvania, but state officials refused to participate in the scheme.

Meanwhile, allies of Trump proposed using armed private contractors to seize voting machines after the election, according to the *Los Angeles Times*, an executive order that was never enacted. Sham audits of results in key states were conducted. Even eighteen months after the election, Trump supporters in multiple states attempted to seize voting-machine data, chasing evidence of vote-rigging conspiracy-theories, undermining the security of elections they claimed to protect. No such evidence was ever found despite Trump's claims that 2.7 million votes had been somehow altered. Indeed, election officials and most independent, non-partisan observers said the 2020 election was the most secure ever. But for Trump supporters, voting machines became the bogeyman. The machinery of democracy came under as much scrutiny in the twenty-first century as the initial voting machines developed in New York did more than a century before. Ironically, the first voting machines were developed to put an end to the voting abuses of paper ballots, something modern voting machine critics advocate a return to.

Voting requires some method of tamper-proof machinery. The ancient Greeks used different colored pebbles dropped into urns to elect five powerful ephors—second in rank to the king—from the ranks of elected assemblymen. Romans as early as 139 BC used paper ballots. Balls, buttons, and tokens were used to cast votes at various times in different countries. The word *ballot* actually derives from the Italian *ballota*, which means *little ball*. But in the United States, a country practically synonymous with modern democracy, paper ballots ruled the latter half of the nineteenth century, despite being a system that lent itself to abuse. Paper-ballot requirements were issued by respective political parties who could reject votes if they allegedly didn't adhere to design specifications. And while ballot boxes might be seen to be under lock and key, false bottoms could hide a number of ballots. And in contentious elections, many voters didn't like that their preferences were readily identifiable by the shape and color of the ballot in their hand. Many became "vest-pocket" voters who kept their ballots hidden until the moment their vote was cast. In short, there was a lot of room for trickery. Or worse.

Edgar Allan Poe, master writer of the macabre story and resident of New York, disappeared on a trip to Philadelphia only to be later found dead in Baltimore outside a polling station shortly after an election's conclusion. While Poe's death remains a mystery, one theory suggests Poe was kidnapped by a "cooping" gang that forced him to vote multiple times in an election (Hence the phrase *cooped up*.) The practice was widespread in the nineteenth century as was the use of *floaters* who would sell their vote to the highest bidder.

The voting machine that would dominate politics for the next century evolved out of a device used for an election in Lockport, New York, in 1892. Patented by Jacob H. Myers of nearby Rochester, the machine built on an earlier design that used push buttons to vote. "The Myers machine makes it impossible to buy votes with any certainty that the goods will be delivered," reported the *New York Times* in a detailed account on April 13, 1892. "Each voter entered the booth alone and as he passed in his name was announced and recorded by one of the clerks. One minute was allowed each man for voting although few required more than from ten to fifteen seconds to register ballots for between fifteen and twenty candidates. When the voter pressed the knob opposite any candidate's name that knob, with those candidates of the same office of other parties, was immediately locked, the knob he pressed registering and the others not, thus preventing any fraud on the part of the voter."

The *New York Times* also noted the economic benefit. "The savings to the town by using the Myers machine is very great. The number of polling places can be reduced by fully half or more, fewer inspectors are required, and the material expenses are generally lessened." Election results for sixty-four candidates were known inside of ten minutes instead of the previous three hours. Results from states and national elections could be ascertained in thirty minutes, added the report. In what was considered a severe setback for the opposition, a Republican mayor and six of eight Republican aldermen won election.

Two years later, Sylvanus Davis added a party lever and simplified the mechanism. The finishing touches, so to speak, were added by Alfred Gillespie who used a curtain linked to the lever that recorded the vote. Gillespie also introduced the lever by each candidate's name and worked out how to make the machine programmable so it could support races where voters had to select three out of five candidates, for example. In December 1900 the United States Standard Voting Machine Company was formed with the patents of Myers, Davis, and Gillespie all gathered under one corporate roof. Gillespie was named one of the directors.

Success wasn't instantaneous. One reason may have been the machine's relatively high cost at the time: $550 each. A cover story in *Popular Science* magazine in the early 1920s pushed the voting machine into the public's consciousness where it came to symbolize fair elections and progressive reform. For generations of voters to come, voting was a three-step process:

1. Pull the handle to close the curtain around the voting booth.

2. Turn the levers over the name of the candidates or issue chosen to see the "x."

3. Pull the handle back to record your vote and reopen the curtain.

There was a certain physicality to the process that resonated with voters as did the clanging noise of the machine as the vote was recorded. Many left the voting booth with a smile on their face, satisfied with their act of citizenship. Crucially and unlike twenty-first-century elections, there was little criticism of the voting machine itself.

Some of that satisfaction may have been created by the lever-voting machine's reputation for being indestructible. The voting booth itself was a quirky, steampunk marvel that weighed 700 pounds, had 28,000 moving parts, and the curtain put people in mind of an old-fashioned shower.

And while the voting booth was mounted on wheels, they weren't very good.

It took a lot of muscle to move them. Machines were configured to lock out other candidates when one candidate's lever was turned down. At the close of the election, the results were hand copied by the precinct officer, although some machines would eventually be able to make the tally automatically. Each machine could register 999 votes, a technical limitation that dictated the number of voting districts required. In New York's Erie County, for example, 1200 voting districts were required. By contrast, electronic voting systems reduced the number to 837. Gillespie's company held a virtual monopoly on voting machines until Samuel and Ransom Shoup filed a patent for a competing machine in 1936.

New York loyalists kept using lever machines until the mid-term elections of 2010 when legislation put an end to its use. But the writing was on the wall as early as 1965 when a "Votomatic" punch card system developed by Joseph P. Harris and licensed to IBM arrived in polling stations

By 1996, punch cards were used by roughly a third of American voters. Punch-card voting gained notoriety, however, in the 2002 presidential election when the status of "hanging chads" dangling from punch cards drew scrutiny in the state of Florida. A Supreme Court ruling determined the outcome in favor of Republican George Bush over Democrat Al Gore. Typically, punch-card systems used a butterfly ballot with candidates' names listed in columns separated by a middle column of punch holes read by an electronic reader. Sometimes, the holes weren't punched all the way through, leaving "hanging chads." The Help America Vote Act of 2002 effectively put an end to punch-card voting, with their last use in 2014. Since then, optical scanning machines were tried but finally gave way to direct-recording electronic (DRE) devices ideally backed up by a paper audit trail. DRE machines typically record a vote selected via a touch-sensitive screen or optically scan a paper ballot. With the exception of some small jurisdictions, the use of electronic voting systems with a paper record were essentially universal across the United States.

For most experts, that paper trail is crucial to maintain the integrity of voting machines due to the fear that purely electronic machines could be hacked, and votes altered, a concern that rippled through the 2016 elections and led to voting systems being declared critical national infrastructure in 2017. As a rule, voting machines are not connected to the internet and there is no evidence that remote electronic tampering occurred. By 2020, a new gold standard in vote certification called "risk-limiting audits" (RLA) was in place to make sure the announced winner was indeed that. DRE machines can malfunction if they are not replaced in a timely fashion. A few instances in which votes were "flipped" by old malfunctioning machines with outdated software were magnified by conspiracy theorists. Chief among them was President Trump, who repeatedly claimed that 2020 presidential elections were somehow rigged. Multiple, extensive, and expensive audits of election results recorded by machines like those made by Dominion Voting Systems Corporation failed to find any evidence of voter fraud that would have altered the election results.

A newer approach to voting is ballot marking devices (BMD) that allow voters to vote using a computerized interface by following visual or audio prompts. And unlike the lever machines that seemingly lasted forever, electronic voting machines have an expected lifespan of ten to twenty years, according to the Brennan Center for Justice, with replacement parts perhaps becoming unavailable even sooner. Costs associated with voting machines—including software updates, security patches, and licensing fees—are now measured in the hundreds of millions of dollars. The price tag is likely to grow as new technologies arrive in the future.

Nevertheless, with every election going forward, the voting machine itself is going on trial. In a 2022 primary election in Otero County,

New Mexico, Republican officials refused to certify the election over concern that the voting machines were somehow compromised. These fears, the legacy of Trump's "Big Lie," were unfounded and certification finally came at the behest of the New Mexico's Supreme Court. Many observers fear that every election will be challenged in this way and slowly undermine the faith people should have in the election process. Some election experts like Eddie Perez, speaking to the *Associated Press* in November 2022, of the non-partisan OSET Institute think continued criticism of voting machines is a "manufactured subversion" designed to sow mistrust and to set the stage for election results to be controlled by partisan state legislatures rather than by voters. Ironically, some people advocate a return to the paper ballot, a notion that would return elections back to a time even before the invention of the voting lever machine. Most professionals view the paper ballot notion as wildly impractical when millions of votes need to be counted.

"Machine counting is generally twice as accurate as hand-counting and is a much simpler and faster process," Stephen Ansolabehere, a professor of government at Harvard University, told the *Associated Press* in November 2022. In one New Hampshire study, hand-counting was off eight percent of the time while machine counting had an error rate of a half-percent, said Ansolabehere.

Experts believe that the use of paper ballots would be open to abuse and tempt partisan actors to issue premature victory claims as hand-counting could take weeks. And whether hand-counting would be trusted is an open question. That kind of abuse of the voting process is the very thing the New York inventors of the mechanical-lever voting machine over a century ago were trying to eliminate.

9

Tesla Invents the Remote Control (1898)

Figure 9. Nikola Tesla, c. 1890. *Source*: Library of Congress, Prints and Photographs Division, LC-DIG-ggbain-04851.

If there is a single inventor that could be said to have a cult following it is Nikola Tesla. While figures like Albert Einstein, Leonardo da Vinci, and Thomas Edison are well-known historical personages whose lives are the stuff of school lessons, Tesla lurks in the murky shadows of mysterious inspiration, not quite understood but revered by those with a similar bent. Foremost among them perhaps is Elon Musk who named his electric car after him, putting the Tesla name in front of millions.

Tesla, a Serbian immigrant, lived in New York City for sixty years and there was rarely a dull moment. Much of his pioneering work in the nineteenth century is intrinsic to modern life. These include neon and fluorescent lights, laser beams, X-rays, wireless transmission, an induction coil widely used in radio, and many more, with over three hundred patents to his name. Tesla also played a central role is the "current war" with Thomas Edison, which made alternating current (AC) the electrical system used today—Tesla zapped himself with a 250,000-volt shock to prove its safety. But the Tesla invention many people around the world hold in the hand daily may be his legacy. That's the remote control.

Tesla was born in the town of Smiljan in what is now Croatia but back in 1856 was part of the Austrian Empire. One of five children, Tesla's parents were of Serbian stock—the father was an Orthodox priest and his mother was a stay-at-home mom with a knack for memorizing epic Serbian poems and inventing small household devices in her spare time. Tesla always credited her as the source of his inspiration and his photographic memory. After completing high school, Tesla contracted cholera and was bedridden for nine months. Tesla then evaded conscription into the Austria-Hungarian army, wandering in the mountains, a sojourn Tesla later said made him stronger both physically and mentally. He had grown to become a slim 6'4" figure with arresting eyes.

Tesla's subsequent academic career in Graz and Prague was spotty but by 1882, Tesla had found his footing with a job at the Continental Edison Company in Paris where he gained a lot of practical experience in electrical engineering. Tesla soon came to the attention of his superiors and at the age of twenty-eight was transferred to the United States with a letter of recommendation to Thomas Edison himself from his French boss. It read: "My dear Edison: I know two great men and you are one of them. The other is this young man!"

Tesla already had an idea for an alternating current motor, a notion that he had been turning over in his mind for many years. Edison, however, was wed to direct current and viewed AC as competition. Edison hired Tesla but put him to work on improving his DC generation plants. Edison promised Tesla a $50,000 bonus if he succeeded. After several months work, Tesla succeeded and claimed his reward.

But Edison stiffed him. Edison said the reward offer had been made in jest. "When you become a full-fledged American, you will appreciate an American joke," said Edison. Tesla immediately resigned. Edison would come to rue the day.

With no money, Tesla was forced to take a job digging ditches for $2 per day to survive. But in time, word soon spread that a major talent was languishing in a dead-end job. A group of investors backed the creation of the Tesla Electric Company and while Tesla designed a beautiful arc lamp, he didn't profit from the work. But it got him noticed and another investor agreed to back Tesla's plan for an AC motor.

In 1887, Tesla filed a series of patents for a complete system of AC motors and power transmission. Industrialist George Westinghouse bought the patents for $60,000, kicking off the current war in earnest. Edison went to great lengths to disparage AC, claiming it was unsafe. Edison electrocuted animals using AC power to put a scare into people. An Edison supporter arranged for the first use of an electric chair on a condemned murderer using a Tesla-Westinghouse motor—the prisoner died horribly. But DC had a major drawback: it could only travel about a mile before losing its potency. That meant generators needed to be built in every neighborhood at great expense. Perhaps the most famous of these DC generating stations was the first one Edison built in 1882 at 257 Pearl Street to supply electricity to a few hundred customers in lower Manhattan. Edison essentially believed that a DC generator would become a luxury home appliance while Tesla had a more democratic vision.

Tesla and Westinghouse made their case at the 1893 World's Fair, also known as the Columbian Exposition, in Chicago. The big appeal of AC power was that it could be transmitted hundreds of miles from a central generating station. Tesla warmed up the crowd by displaying the first neon signs, which used what became known as the Tesla coil, that allowed the use of high frequency electricity as a power source.

Tesla then demonstrated the safety of AC by zapping himself with 250,000 volts that danced harmlessly across his body—thanks to rubber shoes and a high-frequency current—and bathed him in a blue halo of electric flame. President Grover Cleveland pushed a button and incandescent lamps turned the 686-acre fairground into a city of light. The Exposition would be nicknamed the "White City" for both the 100,000 light bulbs that allowed fair goers to visit at night and the temporary neoclassical buildings made of staff—a mixture of plaster, concrete, and jute fiber—covered in stucco and painted white to resemble marble. For the twenty-seven million people who visited the fair and strolled through the Great Hall of Electricity, it was clear the future was to be lit by AC.

Tesla fulfilled a long-time dream with the construction of hydroelectric power station at Niagara Falls—eventually power lines from the plant would light up Broadway. The current war, however, had been costly. Westinghouse found himself in financial difficulties and threatened by a horde of robber barons led by financier J. P. Morgan. In a magnanimous gesture, Tesla tore up the contract that gave him generous royalty payments, saving Westinghouse. Tesla was grateful to the one man who had believed in his invention, but Tesla would experience financial difficulties for the rest of his life.

Tesla became a celebrity scientist. A new lab was built in a factory building on what is now called LaGuardia Place. He was a regular at Delmonico's, the best restaurant in Manhattan and the hub of New York society. A reporter for the *New York World* described Tesla as "almost the tallest, almost the thinnest, and certainly the most serious man who goes to Delmonico's regularly."

Tesla moved into the posh Gerlach Hotel on Twenty-Seventh Street, but his favorite hangout was the exclusive Players Club in a Gramercy Park townhouse where artists mingled with businessmen and the famous writer Mark Twain was counted as a member and friend to Tesla. The Players Club was started by Edwin Booth, Shakespearean actor and older brother of John Wilkes Booth, President Lincoln's assassin, who hoped to repair the family name.

Disaster struck in 1895 when a lab fire destroyed all of Tesla's work. Tesla fell into a deep depression, but The Players Club organized a benefit and this along with self-prescribed electroshock treatments, soon had Tesla on his feet and in a new lab at 46 East Houston Street. At this lab, Tesla investigated X-ray images he called "shadowgraphs" and wireless transmission.

Tesla's big comeback was at the Electrical Exposition of 1898 in Madison Square Garden. Tesla demonstrated a device "that caused a sensation such as no other invention of mine has ever produced," he later wrote. It was the first demonstration of a remote control, and it was beyond the limit of known technology at the time. So much so that examiners from the United States Patent Office thought the claims made in the application "Method of and Apparatus for Controlling Mechanisms of Moving Vessels or Vehicles" improbable until the actual demonstration.

Tesla used the first remote control to control a battery-powered, four-foot-long model torpedo boat. The key element was a new kind of radio-activated switch that allowed Tesla to control the boat from afar thanks to a boat deck studded with antennas and a control box for navigation featuring levers and a telegraph key

The crowd witnessing the demonstration was stunned, with some believing the craft was operated via using mind control as radio waves were still a relatively unknown phenomenon. Others believed it was magic. Tesla, meanwhile, had some fun at the crowd's expense, letting them think they could control the boat with shouted commands. "What is the cube root of 64?" one asked. Everyone gasped when the boat's lights flicked on and off four times.

Newspapers were agog and focused its possible use on a wirelessly controlled torpedo. Tesla had one-upped Edison again. Edison, the wizard of Menlo Park (across the Hudson River in New Jersey) had been involved in the development of a wired torpedo in 1892. Mark Twain volunteered to represent Tesla in the sale of this "destructive terror which you have been inventing" in England and Germany. Ever the visionary, Tesla told a *New York Times* interviewer: "You do not see there a wireless torpedo; you see there the first of a race of robots, mechanical men which will do the laborious work of the human race." The remark basically anticipated drones.

Financially, Tesla's remote control was a flop. It was so far ahead of its time. Investors of the period didn't see it as having any commercial potential. The intended client, the United States Navy, rejected the technology as being too flimsy. That was a shortsighted assessment. During World War I, the German Navy experimented with remote control guided boats filled with explosives and by World War II, navies on both sides of the conflict were using remote-controlled torpedoes and missiles. Tesla would lament that his remote control became popular long after his patent expired and was of no benefit to him financially.

Garage-door openers, radios, and other devices were soon equipped with remote controls, but remote-control use only began to soar when it was married to the television in the 1950s by Zenith engineers. The company's "Space Command" control used ultrasonic frequencies that were inaudible to humans but that were the bane of dogs everywhere. To the relief of hounds, remotes in the 1980s began using the infrared light signals commonly used today. The couch potato was born.

Remote control swiftly became a feature of many devices, with the circuitry small enough to fit inside a smartphone. Remote control tech cemented its place in modern history when it left Earth, controlling the Mars Pathfinder in 1997 as well as many extraterrestrial vehicles and orbiting satellites since then. On the home front, the remote control continues to be linked to a seemingly infinite number of devices ranging from air conditioners to cameras. Wireless charging, once considered a white whale, is now used to recharge cell phones.

Tesla, however, became mired in difficulty. Tesla decamped to Colorado Springs where he continued his research into wireless transmissions, succeeding in creating man-made lightning and illuminating two hundred lamps from twenty-five miles away. Tesla also was certain that he had received radio signals from outer space, a claim met with derision at the time, but which proved to be correct. Returning to New York and new rooms at the Waldorf-Astoria Hotel, Tesla began construction of a wireless transmission tower on Long Island that would broadcast news, music, and stock reports around the world. "When wireless is fully applied the earth will be turned into a huge brain," said Tesla.

However, a financial panic, labor issues, and the withdrawal of support from J. P. Morgan ended the project. Then in 1901, Guglielmo Marconi sent a radio signal transmitted from England to Newfoundland. Making matters worse was that several patents held by Tesla and used by Marconi were later controversially revoked in favor of the Italian, depriving Tesla of much needed income. The United States Supreme Court would revoke that decision after Tesla's death.

In 1905, Tesla suffered a nervous breakdown. In 1917, Tesla received the Edison Medal from the American Institute of Electrical Engineers but was disappointed that reports of his winning the Nobel Prize proved erroneous. In 1919, Tesla suggested ground-based wireless transmissions could be used to fly supersonic airships. But over the ensuing decades, Tesla's invention career stalled. Tesla maintained a modest consulting engineer's office at 8 West Fortieth Street and was well known for feeding the pigeons in nearby Bryant Park.

But the ideas kept coming. In 1932, Tesla envisaged a "thought" camera that could record mental images that he believed also appeared on the retina of the eye. At the age of seventy-two, Tesla filed a patent for an early version of a vertical takeoff and landing (VTOL) aircraft but never had the funds to develop a prototype. Electric VTOLs are now considered to be the future of short-flight aviation. The intersection of West Fortieth Street and Sixth Avenue was later designated Nikola Tesla Corner by the city.

Tesla never married and always lived in hotels, residing at the Hotel New Yorker in his later days. Tesla grew increasingly eccentric, insisting that people keep three feet away to avoid their germs. In 1934, Tesla claimed to be developing a remotely controlled "death ray" or particle beam that could shoot down airplanes from 250 miles away, a notion that generated much interest as World War II loomed. In 1939, the Soviet Union paid Tesla $25,000 to investigate beam weapons. But as the conflict began, Tesla's health deteriorated, made worse by injuries suffered in 1937 when he was struck by a taxi but for which he didn't seek treatment. As the war began, Tesla appeared deathly skinny and was reportedly prone to fainting.

Tesla died in his sleep on January 7, 1943, discovered by a cleaning maid in his rooms (3327 and 3328) on the thirty-third floor of the Hotel New Yorker. Tesla was eighty-six years old. His death created a massive scramble, according to the Science History Institute. No one was sure whether the death ray was real. Tesla's nephew, Sava Kosanovic, was a Yugoslavian diplomat based in New York. Upon hearing of his uncle's demise, he rushed over to the hotel, ordered a locksmith to crack open Tesla's safe, and left with a memorial book from Tesla's seventy-fifth birthday. The United States government feared Kosanovic was a spy. Kosanovic maintained that someone had already gone through Tesla's papers before his arrival and that technical papers and a large notebook with some pages marked "Government" were missing.

Despite Tesla being an American citizen since 1891, the Office of Alien Property Custodian swiftly impounded all of Tesla's belongings. A physics professor and inventor from the Massachusetts Institute of Technology (MIT) and linked to the government's Office of Scientific Research and Development was tasked with examining Tesla's papers to see if he had indeed harnessed "the bolt of Thor." After three days of study, the MIT expert decided Tesla's death ray research was speculative at best. A supposed prototype stored in a vault at another hotel proved to be nothing of value. In 1952, the United States courts awarded Kosanovic possession of his uncle's papers, a move that gave Soviet scientists access when they were transferred to a Belgrade museum.

Over the course of the next twenty years, the Soviets teased that a fantastic weapon was in development. A report in the highly respected *Aviation Week & Technology* magazine showed leaked diagrams remarkably similar to Tesla's death ray—the United States had kept copies of Tesla's originals, which have since mysteriously disappeared. Compounding the mystery is a 1976 FBI document addressed to then FBI Director Clarence Kelly stating a one-month long analysis of Tesla's papers by scientists of the United States Navy and Office of Strategic Services (OSS) was hardly sufficient. While the author's name is redacted, the document re-assessing the possible military applications of Tesla's papers notes the inventor's "torturous" handwriting was "small, blurred, and as difficult to translate as a foreign language" even though he wrote in English. The same writer suggests that Tesla may have been working to harness ball lightning as an energy source and particle accelerator for his death ray. Tesla had already succeeded in creating man-made lightning while in Colorado.

Then during the Reagan administration, the United States moved to develop its own death ray. Called the Strategic Defense Initiative, a.k.a. Star Wars, it consisted of orbiting satellites equipped with particle beams or lasers that would shoot down enemy missiles. While it never came to fruition, the Star Wars program is credited with contributing to the demise of the Soviet Union who could not match the potential expenditure. Still, the death ray weapon continues to capture the imagination. In 2019, President Donald Trump authorized millions of dollars for research into space-based particle-beam weapons. Remarkably, the MIT physicist that debunked Tesla's death ray work was the president's uncle, John Trump.

Israel, meanwhile, has developed a "laser wall" of ground-based beam weapons that appears to be very much like Tesla's original concept. The "game-changing" Iron Beam defense system can target aircraft, missile, drone, and other aerial threats at a cost of about $3.50 per shot, making it more economical that previous missile defense weapons, said Israeli defense minister Benny Gantz in 2022

It was as if Tesla was remotely controlling events even after his passing. Hundreds attended Tesla's 1943 funeral service at the Cathedral of St. John the Divine. Three Nobel Prize winners addressed mourners, noting the loss of a great genius "who paved the way for many of the technological developments of modern times." They had no idea that Tesla's influence on modern times would be so long lasting.

10

The Cool Machine (1902)

Figure 10. Willis Carrier, 1915.

Willis Carrier was standing on a train platform in 1902 watching the fog roll in when he came up with his cool idea. At that moment, Carrier was focused on how to reduce the temperature inside a printing plant. Little did he realize that the ripple of effects of his air conditioning invention would change the world in ways still being experienced. Air conditioning would change the way people lived, make formerly inhospitable places livable, and alter the political landscape. Thanks to Carrier, for the first time in recorded history, people had a refuge from hot weather. It changed everything.

Carrier was born in 1876 into a family of farmers in Angola, New York, a small village twenty-two miles southwest of Buffalo. As a child, Carrier struggled with the concept of fractions. His mother cut apples into variously sized fractional pieces to help him understand the math. Carrier later said it was the most valuable lesson he ever had as it taught him the value of intelligent problem solving. Sadly, Carrier's mother died when he was only eleven.

Carrier worked his way through Cornell University on a scholarship, graduating in 1901 with a degree in mechanical engineering. Carrier had grown into a 6'6" young man with a robust physique honed by an active involvement in basketball, swimming, skating, and boxing. A month after graduation, Carrier landed a ten-dollar-per-week job with Buffalo Forge Company, a maker of fans, heaters, and industrial drying equipment with roots in blacksmithing. At first, Carrier's insistence at bringing scientific methods to bear on manufacturing caused some to bristle until they realized his methods were saving the company substantial amounts of money. Carrier then became a one-man Department of Experimental Engineering.

The timing was perfect. In Brooklyn, the Sackett-Wilhelms Lithographing and Publishing Company was crying out for help. The summers of 1900 and 1901 were especially hot and humid. High humidity caused the paper to swell, which meant that images were blurred rather than being sharply defined. Paper losses were almost ruinous. With the summer of 1902 warming up, the company contacted Buffalo Forge, hoping to tap into their industrial drying expertise.

Carrier had never even heard of humidity. Prior attempts at air conditioning often involved running a fan over racks of ice carved in the New York area, for example, from a frozen Hudson River in winter and stored in specially designed ice houses. This method used up tons of ice daily and would only lower the temperature by a few degrees if at all. Compounding the problem was a Victorian etiquette that required people to ignore summertime heat. If heat wasn't a problem, then a machine designed to deal with it wasn't necessary. The idea of air conditioning wasn't in most people's brains.

Buffalo Forge dumped the problem into Carrier's lap, and he focused on developing a solution with the single-mindedness that would eventually earn him a reputation for having his head in a cloud of calculations, tales amplified no doubt by his height. Carrier, so it was said, would order a three-course meal and not touch a morsel or pack a suitcase with only a handkerchief inside.

Armed with a slide rule, some temperature-humidity charts from the United States Weather Bureau and the floor plan of the Sackett-Wilhelms building, Carrier explored possible solutions. The breakthrough came in a flash of insight while Carrier was waiting for train in Pittsburgh on a foggy night. Carrier noticed the mist in the air and the condensation on the train windows and realized that the fog itself could be used to gather moisture and make it condense. By creating an artificial fog and controlling the temperature of the air and the temperature of water creating the fog, Carrier could produce any level of humidity at any temperature. A warm fog would create more humid air while a cold fog created drier air. Modern air conditioning was born.

Carrier's notion wasn't immediately embraced. The idea that water could be used to control humidity drew incredulity and ridicule. Undeterred, Carrier cobbled together a prototype machine made from an air washer, miscellaneous Buffalo Forge parts, and nozzles used to spray insecticide. In 1904, Carrier patented his Apparatus for Treating Air.

Carrier's air conditioner, as it became known shortly thereafter, was geared toward manufacturing plants. Things started slowly with just one unit sold in the first year. But word spread and sales gradually increased. In 1907 Buffalo Forge created a subsidiary called the Carrier Air Conditioning Company. In 1911, Carrier went public with his research by presenting a paper called the *Rational Psychrometric Formulae* at a meeting of the American Society of Mechanical Engineers.

What the paper lacked in glamour was more than made up for by its usefulness. The paper provided a series of calculations that would allow any ventilation designer to precisely tailor air conditioning equipment to work with maximum efficiency in any sized location. It was like a Rosetta Stone for engineers.

Then in 1914, in what must rank as one of the most shortsighted moves in business history, Buffalo Forge decided to shut down the Carrier Air Conditioning Company. Rising sales were deemed a threat to Buffalo Forge's other businesses and executives were swept up in a belt-tightening fever as World War I began. Carrier's job would be safe but everyone else would be fired.

In response, Carrier and six other engineers resigned from Buffalo Forge. With a pooled fund of $32,600, Carrier formed the Carrier Engineering Company. Office furniture was secondhand, but Carrier's reputation was now first-rate with forty contracts secured in the first year. Planning for each was meticulous with factors like building size and construction, location, the number of people it occupied as well as the number of hours they would be on premises, and the power requirements of on-site machinery. For Carrier, air conditioning was as much an art as it was a science and his near-obsessive dedication and exuberance got customers excited. Carrier also proved to have a flair for marketing. Mech, the mechanical weatherman, became a cartoon strip superhero, solving businesses' air-conditioning problems with ease. When the United States entered World War I in 1917, Carrier constructed what was believed to be the largest air conditioning plant ever assembled to keep explosive munitions stable.

Still, the idea of air conditioning hadn't taken hold with the average person who for the most part still had a Victorian attitude toward heat despite mounting evidence that heat was not only uncomfortable but also deadly. Air conditioning was still expensive so while it seemed a prudent business expense, it was not yet thought of as a device for personal comfort.

The movies changed all that. When the Great War ended and the Roaring Twenties began, the idea of personal comfort finally entered people's minds. In 1925, Carrier installed an air conditioning system in New York City's massive movie palace, the Rivoli Theater. Construction was slightly delayed when a New York City buildings inspector refused to issue a permit on the grounds that he had never heard of dieline, the refrigerant Carrier was using. Carrier proved that dieline was completely safe by pouring some into a cup and tossing in a lit match. A slightly unnerved inspector issued the permit.

The Rivoli opened on Memorial Day to an eager standing-room only crowd, perhaps more intent on the AC than the film. The system kicked in late, and fans were in every lady's hand. But as the film progressed so did the theater cool and fans were put away. The movie business never looked back as air conditioning became quickly associated with the movie experience. For the first time, the average person had a place to go to escape the summer heat. Other theaters soon followed as did shopping malls. Air conditioning became expected in any large public gathering place. Trains outfitted with air conditioning began to roll and travelers soon expected the same comfort level in their hotel rooms.

Still, it took decades for air conditioning to work its way into every building. Home air conditioning units debuted in the 1930s but were so frightfully expensive that few people in the Depression-era bought them. During World War II, "limitation" orders prohibited the manufacturing of air conditioning units for personal comfort—existing AC units installed in movie houses and department stores were diverted to factories making tanks. Electric fans ruled. In New York City, people continued to sleep on fire escapes to escape sweltering apartments and screened-in porches offered nighttime respite from the summer heat in private homes.

Willis Carrier died in New York City on October 7, 1950, at the age of seventy-three just as a decade of consumerism began that made window air conditioners a status symbol, marketed alongside television as the epitome of modern luxury even as unit prices, driven by more manufacturers, became more affordable. Ironically, Carrier had never gotten around to installing air conditioning in his own home, but millions added air conditioning to their own. Air conditioning became the norm for new skyscrapers, but universal adoption progressed in fits and starts. It wasn't until 1961, for example, that Lincoln Center in New York City became air conditioned, allowing the performing arts center to move from seasonal performances to a fifty-two-week schedule. By the late 1960s, air conditioning became an affordable luxury in cars. In the 1970s, librarians and clerks in the un-air-conditioned branches of the New York Public Library calculated a summer day's temperature-humidity index (THI) to learn whether they would be getting double pay on the job or a single day's wages if they were told to stay home. And there was no guarantee of an air-conditioned ride on New York City's subways until the 1990s.

Willis Carrier left an enduring legacy, much of which he couldn't have foreseen. Air conditioning is arguably responsible for a major shift in population in the United States to the Sunbelt. In 1950, the population of Phoenix, Arizona, was 107,000. In 2021, that number stood at 1.6 million. Likewise, cities like Dallas, Houston, Atlanta, and others likely would not be the major metropolises they are today without air conditioning. That shift in population also changed the political landscape as the Sunbelt states Congressional representation in Washington, DC, increased. Air conditioning makes unlivable places livable.

More controversially, air conditioning also has been fingered for various environmental problems. Spikes in air conditioning use have been blamed for blackouts and brownouts during heat waves. Legionnaire's disease was caused by inhaling bacteria in water droplets from air conditioning systems and it proved deadly. Freon, a refrigerant used for decades, was discovered in the 1970s to be the cause of a thinning ozone layer in the upper atmosphere. While chlorofluorocarbons (CFCs) like Freon were ultimately banned, the effects last decades. Unfortunately, hydrofluorocarbon (HFC), the refrigerant that replaced CFC, traps heat in the atmosphere that would normally dissipate into outer space. The Kigali Amendment to the 1987 Montreal Protocol on ozone pollution, signed by President Joe Biden in 2022, requires signatories like the United States to reduce HFC use by eighty-five percent by 2035.

Alternative "naturally cool" refrigerants like carbon dioxide, used for ice rinks at the 2022 Winter Olympics in China, are being tried.

On the commercial side, a company called Blue Frontier is developing a novel AC machine that uses a liquid to suck moisture out of the air to lower humidity and then uses that water to cool down a room, reducing electricity consumption by fifty to ninety percent depending upon weather conditions. Researchers at Harvard's Wyss Institute are developing an AC unit called coldSnap that uses a special coating on a ceramic frame to evaporate water to cool a room without adding moisture to the air. AC manufacturers, meanwhile, are hoping to extract "greener" performance with more clever designs.

On the plus side, air conditioners have become more energy efficient, especially models with variable speed compressors that allow the machine to run on different settings, and there is more awareness that air conditioners need to become more environmentally friendly. More than ninety percent of American homes have air conditioning. With climate change causing a rise in temperatures, no one is giving up their air conditioning any time soon. In fact, the International Energy Agency estimates the global demand for space cooling will triple by 2050. The paradox is that while air conditioning offers relief from the heat, the power it uses is a significant source of the greenhouse gases causing the temperature rise. Government figures show that air conditioning accounts for one-fifth of total energy use in the United States. Alternative power sources like solar and wind may offer a solution and air conditioning itself may change due to technological advances and alternative refrigerants. That would be cool. Otherwise, we may have to become more comfortable with discomfort. Again.

11

The Teddy Bear Is Born in Brooklyn (1902)

Figure 11. Teddy bears made in New York City. *Source*: Library of Congress, Prints and Photographs Division, LC-USZ62-108039.

Winnie the Pooh, Paddington Bear, the Care Bears, and all the teddy bears in existence today owe a debt of gratitude to an American president and the husband-and-wife duo that created the teddy bear in a Brooklyn candy store in 1902. But the true story of the creation of the teddy bear is more grizzly than cute.

The central figure in the creation of the teddy bear is the man for whom it is named: President Theodore "Teddy" Roosevelt. The twenty-sixth president of the United States assumed the office in 1901 following the assassination of President William McKinley in Buffalo, New York. Roosevelt was hiking Mount Marcy, the highest peak in New York's Adirondack Mountains, when he received the news of McKinley's death. Roosevelt's midnight stagecoach ride to the nearest railway station is the stuff of local legend. That Roosevelt was in the Adirondacks should come as no surprise. Roosevelt was a New Yorker through and through, having been born in a brownstone at number twenty-eight West Twentieth Street in New York City to wealthy parents. Roosevelt was elected to the state legislature as an assemblyman from Manhattan's twenty-first district in 1881, ran unsuccessfully for mayor of New York City in 1885, served as the city's police commissioner from 1885 to 1887, and then became governor of the state in 1898. In between times, Roosevelt served as assistant secretary for the United States Navy and as the commanding officer of the Rough Riders in the Spanish-American War in Cuba in 1898, gaining national fame for a legendary charge up San Juan Hill.

Roosevelt embraced an outdoor lifestyle, becoming well known for his love of big game hunting and, perhaps incongruously to modern eyes, a conservationist. President Roosevelt, in fact, created five national parks, 150 national forests, and numerous other conservation initiatives, including the creation of the United States National Forest Service.

As president, Roosevelt didn't have any easy time of it. In 1902, Roosevelt had mediated a nasty strike by miners against coal operators. So, when Governor Andrew H. Longino of Mississippi invited Roosevelt on a bear-hunting expedition, the president accepted it in the guise of a much-needed vacation. Presidents, however, don't really go on vacation. Roosevelt went to Mississippi to try to stamp out racism in the state.

Governor Longino was the first governor of Mississippi to be elected after the Civil War who was not a Confederate veteran. Longino had ambitions beyond the state border—he was running for the United States Senate while still in the governor's seat. The thorn in Longino's side was an avowed white supremacist named James K. Vardaman who was running for the governor's seat.

Roosevelt knew that Mississippi was a hornet's nest. Vardaman, running as a Democrat in opposition to Roosevelt's Republican Party, fanned the fires of racism and made Roosevelt a target, noting that Roosevelt had met with Booker T. Washington, one of the leading Black activists of the day, in the White House. Roosevelt's "social equality" policy resulted in the appointment of many Blacks to federal offices in the south to the ire of many local whites. For most northerners, noted the historian Eugene E. White, Vardaman was "an accident of the times—a fragment of red-hot lava belched from a hitherto quiescent volcano of race hatred."

Vardaman openly endorsed lynching and restrictions on the education of Blacks. His appeal wasn't universal. The *Biloxi Herald* condemned Vardaman "as the most dangerous man to ever aspire to be governor of the state." But for many poor, rural whites, Vardaman was the great white chief. Roosevelt, no shrinking violet, was shocked at the abusive language Vardaman used to attack him. In a 1903 letter about Vardaman, Roosevelt wrote: "I could not even speak to you the unspeakably foul language which he has used against me. It is the filth which even the foulest New York blackguard would not dare use on the stump nor the foulest New York newspaper print."

It was in this toxic atmosphere that Roosevelt found himself. Both Roosevelt and Longino hoped that the president's visit to Mississippi would help blunt Vardaman's racist campaign. In this attempt, neither man was successful. Vardaman was elected governor and later won a seat in the United States Senate. Longino lost his bid for the Senate and retired from public life. And for Roosevelt, the race issue only got worse. In 1903, Roosevelt closed the post office in Indianola, Mississippi, when locals ousted postmaster Minnie Cox because she was Black. The only positive thing to come out of Roosevelt's trip to Mississippi was the teddy bear.

On a November day in 1902, Roosevelt, already piqued by events in Mississippi, was further exasperated when on the first day of the hunt, no bear had been discovered. "Perhaps they were Democratic bears and took to the woods upon my arrival," quipped Roosevelt. The ten-day bear-hunting expedition, centered in the swamps around the town of

Onward, thirty miles north of Vicksburg, was substantial, involving trappers, horses, fifty hunting dogs, supplies, and a coterie of accompanying journalists. Their guide was a former slave and Confederate soldier with the Ninth Texas Calvary named Holt Collier. The then fifty-six-year-old Collier had a fearsome reputation as a bear hunter—he is credited with killing over three thousand bears, more than legendary frontiersmen Davy Crockett and Daniel Boone combined (or so it was claimed). Bear hunting in the swamps was especially dangerous, a challenge relished by Roosevelt, which is why Longino engaged the services of the experienced Collier. "He was safer with me than all the policemen in Washington," Collier would later say.

Collier, eager to please an impatient president, went out early on the second day of the hunt along with some of the hunting dogs. Collier soon cornered a large, male bear that promptly crushed one of the hunting dogs to death. Wanting to save the kill for the president, Collier swung his rifle, smashed the bear in the skull and tied it to a willow tree. When the Roosevelt arrived, however, he was appalled by the scene. Despite being urged to shoot the bear, Roosevelt refused, saying that such a kill would be unsportsmanlike. Collier would kill the bear himself and the animal was slung over a horse and brought back to camp.

News of Roosevelt's compassionate gesture quickly spread. Political cartoonist Clifford K. Berryman immortalized the moment in a sketch in the *Washington Post* soon afterward. Berryman was the foremost political cartoonist of his day—a sketch in 1898 called "Remember the Maine" became the Spanish-American War's battle cry—and would later win a Pulitzer Prize. In Berryman's "Drawing the Line in Mississippi," Roosevelt appears in his Rough Rider uniform with a rifle in one hand and an outstretched hand toward a bear cub with a rope around its neck. Roosevelt is refusing to shoot the bear cub. The cartoon was seen as a double entendre reflecting Roosevelt's opposition to lynching in the South.

The drawing drew widespread attention. Among those who noticed the cartoon were Morris and Rose Michtom, a married Jewish couple who had emigrated from Russia and owned a small candy store on Tompkins Avenue in the Bedford-Stuyvesant section of Brooklyn. Rose fashioned a piece of plush velvet, sewed on some eyes, and displayed "Teddy's bear" in their shop window. Customers were soon asking to buy the bear. The Michtoms assumed they would need White House permission to sell the bear so they mailed the original to the president as a gift to his children and asked if Roosevelt minded if they used his name on the bear. Roosevelt consented and thus the teddy bear had its name. The teddy bear became so popular that Roosevelt adopted it as the mascot of the Republican Party for the 1904 election. Adult supporters of Roosevelt were seen around town clutching their teddy bears. The teddy bear moniker got a boost in popularity in 1905 with a popular newspaper serial about the adventures of the friendly Roosevelt Bears, Teddy B. and Teddy G., written by Seymour Eaton

For the Michtoms, the teddy bear was a gold mine. They got out of the candy business and started the Ideal Novelty and Toy Company to manufacture teddy bears. The early plumpish teddy bears had long arms, ears set far apart, a wide forehead, a stitched nose that gave the head a triangular shape, slightly pointed paws, and plush mohair coats. Ideal would become a major toy maker over the ensuing decades until its acquisition by CBS in the 1980s.

Unbeknown to the Michtoms, in Germany Margaret Steiff had moved from making stuffed elephants to stuffed bears. An American buyer placed a large order, but those initial bears were "lost at sea." Steiff nevertheless went on to become a major manufacturer and adopted the name teddy bear as well. Steiff would sell 975,000 teddy bears in 1907.

Other teddy bear manufacturers jumped in, and teddy bears were made internationally. One of the earliest was Gund, which in 1948 won a toy-making Disney license that included Winnie the Pooh. Gund, in partnership with Hanna-Barbera, won the rights to Yogi Bear, a popular cartoon character said to be named after the great New York Yankee baseball player Yogi Berra. Another popular incarnation was Build-A-Bear Workshop founded in the mid-1990s, which allowed children to supervise the construction of their own teddy bear, a marketing scheme that's grown to include "naughty" teddy bears aimed at adults. In modern times, the Vermont Teddy Bear Company, for example, sells about

500,000 teddy bears per year. Teddy bears have found their way into the classroom as well. In Ireland, for example, medical students visit schools to diagnose "sick" teddy bears with the goal of making children more comfortable around doctors and hospitals. The teddy bear has never lost its popularity. Ironically, Roosevelt never liked his "Teddy" nickname and asked close friends to call him Theodore.

12

The Teenage Debutante Who Let Women
Breathe Easier (1910)

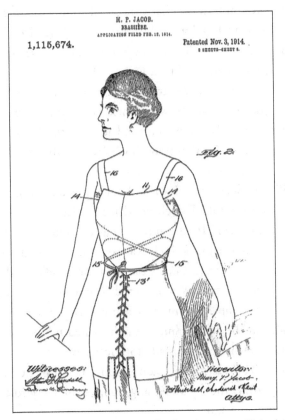

Figure 12. Phelps's 1914 patent drawing for her brassiere. United States Patent and Trademark Office, US Patent 1,115,674, filed February 12, 1914, and issued November 3, 1914.

When Mary Phelps Jacobs invented the brassiere—shortened to bra over the years—it was a shocking sensation. It was probably the least scandalous thing she ever did. But it took a world war for it to catch on.

Women have sought support since at least the times of the ancient Greeks when women wrapped a band of wool or linen across their breasts, pinning or tying it in the back. Bikini-like garments were a feature of Minoan civilization 4,500 years ago—the "Snake Goddess" of fertility is portrayed wearing the "Minoan bra" that emphasized breasts in paintings and figurines discovered among various archeological sites around Crete. By contrast, ancient Roman and Greek women used bra-like straps to reduce their apparent bust size. Early medieval chroniclers mention "breast bags" or "shirts with bags" worn by women. In 2012, lace fragments of underclothing dated to the fifteenth century were discovered in Lengberg Castle in Austria—styles included a short, high-necked sleeveless fabric stretching over cups to cover the cleavage, a variant using two broad shoulder straps, and another with linen cups with a linen extension fastened at the side. The apple-breasted look of the Middle Ages went out of style, replaced by the mono-boob look of the corset.

While the inventor of the corset is unknown, they first appeared around 1500 in Spain and Italy, although the term *corset* appears to be derived from the Old French word for bodice. When Catherine de Médicis married Henri II, the future king of France, she brought the corset with her and banned thick waists at court. For the next several centuries, corsets became a must-wear item for middle- and upper-class women in Western society. Women tried to mirror the "ideal" image of a woman with an ample bosom atop a tiny waist. Some nineteenth-century adaptions of the corset almost foreshadowed modern ladies' undergarments. One version added pads to boost cleavage while the "corslet gorge" designed by Herminie Cadolle split the garment in two, with the top supporting the breasts using straps while a girdle-like corset wrapped around the waist.

Revolution was in the air, however. Corsets made from whalebone and steel rods were designed to narrow an adult woman's waist to thirteen inches and sometimes as small as ten inches. Doctors were becoming increasingly concerned about how the corset artificially restricted the movement and breathing of women to the point of fainting, not to mention damaging internal organs. Organizations like the Rational Dress Society and the Dress Reform Association believed the greater participation of women in society would require emancipation from the corset.

A nineteen-year-old New York socialite changed everything. Mary Phelps Jacobs—"Polly" to her friends—was born on April 20, 1891, in New Rochelle, New York, with steamboat inventor Robert Fulton counted as among her notable American blue-blooded ancestors. Jacobs herself later noted that she grew up "in a world where only good smells existed." At nineteen, Jacobs was making the rounds of New York high society's debutante balls. One evening in 1910 and with another ball just hours away, Jacobs was exasperated. She planned to wear a sheer evening gown to a debutante ball, but the norms of the time meant that she needed a bulky corset to support her ample bosom. When she tried on the dress, she discovered that the whalebones poked out visibly around the plunging neckline and were visible under the sheer fabric. The corset, she wrote later in her autobiography *The Passionate Years*, was "a boxlike armor of whalebone and pink cordage." Her solution was simple, ingenious, and changed women's wear forever.

"Bring me two of my pocket handkerchiefs and some pink ribbon," Jacobs instructed her French maid Marie who helped her sew the material into a simple brassiere. Jacobs wore her creation to the ball and her invention became the talk of the party. Other girls wanted to know how she moved and danced so freely, then asking Jacobs to sew bras for them too. When strangers began offering her a dollar for one, Jacobs realized she had a business. A patent for her "backless brassiere" was issued in 1914 Jacobs noted in her application that her invention was "well-adapted for women of different size" and "was so efficient that it may be worn by persons engaged in violent exercise like tennis." The brassiere's benefits, she noted, "may be summarized by saying it does not confine the person anywhere except where it's needed." The word "brassiere" was about to enter common usage.

Jacobs started her own company and while she landed a few orders from department stores, her "backless brassiere" failed to attract the female public's interest. At the urging of her new husband, Jacobs, now married into the wealthy Peabody family of Boston, sold her patent to the Warner Brothers Corset Company in Connecticut for fifteen hundred dollars.

As it turned out, the sale was ill-timed. The United States was soon embroiled in World War I and in 1917, the United States War Industries Board asked women to stop buying corsets as the frames were made mostly of metal needed for ammunition and other military supplies. The request freed up wartime steel and women at the same time. The corset ban allowed woman to take on more physically demanding jobs like those in factories. By war's end, the popularity of the corset had faded. Over the next thirty years, Warner Brothers Corset Company would reap in an estimated fifteen million dollars in bra sales. "I can't say the brassiere will ever take as great a place in history as the steamboat," Jacobs wrote. "But I did invent it."

The war upended Jacobs's life as well. In 1920, she fell in love with Harry Crosby, a returning soldier cited for bravery, wealthy scion of a prominent Boston family, nephew of financier J. P. Morgan, and seven years her junior. The blonde, slightly built, charismatic Crosby reportedly wooed her in the Tunnel of Love ride at an Independence Day fair where she was acting as a chaperone. After two scandalizing years, Jacobs's husband Richard Peabody, himself a damaged veteran with a strange obsession with fire engines and who had been drying out in a sanitarium due to drinking problems (one of many visits) during the Tunnel of Love episode, agreed to a divorce and Jacobs decamped to Paris with Crosby, her two children in tow. "For the first time in my life," she wrote after spending the night with Crosby at New York's Belmont Hotel. "I knew myself to be a person." This person needed a new moniker. After considering the name Clytoris, she settled on Caresse Crosby.

Harry Crosby landed a short-lived job at J. P. Morgan before the couple decided to lead "a mad and extravagant life," wrote Harry to his parents, requesting the sale of ten thousand dollars in stock to fund this expatriate lifestyle. A list of friends was a celebrity roster of Bohemian Paris in the Roaring Twenties: Salvador Dali, James Joyce, Ernest Hemingway, Gertrude Stein, John Dos Passos, D. H. Lawrence, Anaïs Nin, Henri Cartier-Bresson, and many others. Caresse achieved notoriety for arriving at the annual l'Académie des Beaux Arts ball topless in a turquoise wig riding an elephant. The couple founded the poetry-oriented, highly respected Black Sun Press and counted James Joyce, Kay Boyle, Ezra Pound, and Hart Crane on their authors list. "Black was Harry's favorite color, and he worshipped the sun," explained Caresse.

Caresse became the literary godmother to a lost generation of expatriate writers in Paris. Other works included the reprinting of the *Hindu Love Manual* found on a trip to Damascus. The pair hosted dinner parties from a giant bed in their townhouse, invited guests to enjoy their huge, marble bathtub together and played drunken polo on donkeys. In 1927, Harry and fellow veteran Ernest Hemingway traveled to Pamplona for the running of the bulls. "H could drink us under the table," wrote Hemingway, who had written *The Sun Also Rises*, a novel about the bulls of Pamplona, the previous year.

Of Caresse, an infatuated Anaïs Nin described her as "a pollen carrier who mixed, stirred, brewed, and concocted friendships." Nin would later write: "Caresse enters with the buoyancy of a powder puff, caressing voice (was this how she gained the nickname from Harry Crosby?), her fur hat, her eyelashes, her smile all glittery with animation. The word on her lips is always yes and all her being says yes, yes, yes to all that is happening and all that is offered her."

Harry and Caresse practiced an open marriage, often fueled by drugs and alcohol, that ended tragically. Harry Crosby shot himself and his twenty-year-old mistress, the wild Boston-born "Fire Princess" Josephine Rotch Bigelow, in a 1929 murder/suicide at Hotel Des Artistes in New York. Harry was found with tattooed suns on the soles of his feet and ochre-painted toenails. He was thirty-one. It's perhaps no coincidence that Harry's death came six weeks after the great stock market crash of 1929. Stock sales had funded his lifestyle for all those years. Harry Crosby became a symbol of the rise and fall of the jazz age in many eyes.

Caresse expanded her publishing business with Crosby Continental Editions paperbacks and published works by Ernest Hemingway and Dorothy Parker. A Salvador Dali portrait painted in 1934 is believed to be of Caresse as they were good friends. As World War II approached, Caresse returned to the United States where she started a magazine called *Portfolio* with luminaries like Henri Matisse, Leo Tolstoy, Pablo Picasso, Albert Camus, and Jean-Paul Sartre among the many contributors. Caresse opened an art gallery in Washington, DC, bought a plantation in Virginia and a sprawling apartment on East Fifty-Fourth Street in New York City, allegedly wrote pornography with authors Henry Miller and Anaïs Nin, and at the age of forty-seven, married a football player sixteen years her junior (it didn't work out). Caresse later carried on a long-term love affair with Black boxer, star actor, and civil rights activist Canada Lee, despite the threat of miscegenation laws that would criminalize the relationship and force them to keep their romance a secret.

Meanwhile, the one-size-fits-all brassiere evolved in the 1920s into an A-, B-, C-, D-cup-sized system, an innovation controversially credited to either William and Ida Rosenthal who founded Maidenform or rival S. H. Camp & Co. In the 1930s, bras took off thanks to innovations like the elastic strap and padded cups. Frederick Mellinger, founder of Frederick's of Hollywood, added the first hook bra and the push-up bra. Decades later, three friends—Lisa Lindahl, and costume designers Polly Smith and Hinda Miller—in 1978 would create the first "Jogbra" sports bra, originally made from two jockstraps. "No man-made sporting bra can touch it," was the product's slogan. In 1990, their Jogbra company was sold to Playtex as competition from Nike and others intensified. Title IX legislation in the United States five years before increased women's participation in sports; the Jogbra and its descendants made participation comfortable. All three women were inducted into the United States National Inventors Hall of Fame in 2022.

Meanwhile, bra development marches on. In 2022, the United States military announced it would develop an "Army Tactical Brassiere" (battle bra) that would be integrated into body armor. Among the considerations for the ATB are the inclusion of flame-retardant fabrics, layered compression, plus structural and protective materials. The battle bra also would be designed for accurate sizing, reliable comfort, moisture management, and breathability.

The battle bra is practically the polar opposite of the "Ban the Bra" movement sparked by feminists in the 1960s but the undergarment's hold on women's fashion nonetheless remains firm. Brassieres are today worn by most women in Western countries. While Caresse Crosby never benefited financially, she no doubt found the evolution of her invention uplifting. She passed away due to pneumonia related to heart disease in 1970 at the age of seventy-eight in a fifteenth-century Italian castle called Castello diRocca Sinibalda north of Rome that she had purchased to start an artist's colony. In 1962, filmmaker Robert Snyder made a documentary about the castle, with Caresse leading viewers on a tour. At one point in the film, called *Always Yes, Caresse*, she pulls down her dress to reveal her ample bosom. She doesn't appear to be wearing a bra.

13

The Man Who Streamlined America (1934)

Figure 13. Raymond Loewy with a model of his Imperial House II. *Source*: Library of Congress, Prints and Photographs Division, Gottscho-Schleisner Collection, LC-G613-74809.

R aymond Loewy is the man who streamlined America and became the father of industrial design because his product designs made life easier for people for decades to come. Loewy's designs became iconic. Logos Loewy-designed became the identifiers for major brands like Coca-Cola, Shell, Exxon, the United States Postal Service, and many others. Trains and cars bore his design stamp and he even developed the livery for Air Force One, the plane used by the president of the United States. During his decades-long career, Loewy designed how America looked.

Loewy was born in Paris on November 5, 1893, and arrived in New York City in 1919, a city where he would rise to fame. His character, however, was forged in France. Loewy was the son of a Jewish father and a Catholic mother. As a child, Loewy lived a middle-class life and was well educated. His father was a business journalist and as a teen, Loewy published a newspaper for his neighborhood. Loewy exhibited a talent for art and a love of trains, planes, and automobiles. At fifteen, Loewy patented the design of a model airplane and created a company to sell them. His entrepreneurial streak became well established with these ventures and he cemented his education with an engineering degree from the University of Paris. Loewy also was mindful of his mother's mantra: "It is better to be envied than pitied."

As it did for millions, World War I altered the trajectory of Loewy's life. Loewy served four years and two months in the French army, enlisting as a private, ultimately attaining the rank of captain. Loewy was wounded in combat—burned by mustard gas—and was decorated for bravery three times. The design impulse never left him, and his dugout was decorated with flowered wallpaper, draperies, and tufted pillows. By the war's end, however, both of Loewy's parents were among the millions who had died of the Spanish Flu. His parents left little in the way of inheritance and two older brothers already had left for America. Loewy followed them wearing his only set of clothes—his army uniform—and forty dollars in his pocket. On the voyage aboard the SS *France*, Loewy passed the time drawing sketches of the ship and his fellow passengers.

Loewy's shipboard sketches caught the eye of the right people—magazine editors and advertising people. While initially hoping to land a job as an engineer, Loewy sensed an opportunity to turn his talent for sketching into a career. A fashion editor at *Harper's Bazaar* took Loewy under her wing and introduced him to the city's elite who were charmed by the French former captain. By 1927, Loewy had an impressive list of clients, including the prestigious *Saks Fifth Avenue* store for which he designed print ads and staff uniforms. His career took off.

Loewy, however, didn't like what he saw in America. "Prosperity was at its peak but America was turning out mountains of ugly, sleazy junk," Loewy later told the *New York Times*. "I was offended that my adopted country was swamping the world with so much junk." Refrigerators had spindly legs or were topped with towering tanks, Loewy observed, while typewriters were enormous and sinister-looking and telephones looked disconnected. "I felt that the smart manufacturer who would build a well-designed product at a competitive price would have a clear advantage over the rest of the field when things would become tough," Loewy later wrote.

No one agreed with him. Loewy toured the American Midwest, visiting city after city and company after company and receiving rejection after rejection of his vision of a better-looking world. "No one in the manufacturing world had ever heard of industrial design and no one was interested," wrote Loewy. "My life was a dreary chain of calls on bored listeners."

Loewy's big break came in London where he was on assignment for a big ad agency. A chance encounter with British and American manufacturer of duplicating machines, an early type of copier, gave Loewy the opportunity he was waiting for. Working over three days with mounds of modeling clay, Loewy designed a machine that transformed a piece of industrial machinery into an item that looked like office furniture—it was immediately put into production. Loewy's career as an industrial designer had begun. The stock market crash of 1929–30 put a dent in Loewy's finances, but he continued to believe in himself. Loewy married in 1931, a relationship that ended in an amicable divorce fourteen years later.

New commissions began to roll in. No item was too small for Loewy's notice, and he even designed a pencil sharpener. Among the most significant items was the 1934 Aerodynamic Hupmobile, a four-door sedan that featured faired-in headlamps, a three-piece windshield, and a tire-carrying fastback rear design. The redesign of the Sears & Roebuck

Coldspot refrigerator that same year cemented Loewy's reputation as sales soared. "What I instinctively believed was being proved by hard sales figures," Loewy later noted. "You take two products with the same function, the same quality, and the same price: the better looking one will outsell the other."

Loewy formulated his design philosophy into a simple principle called MAYA (Most Advanced Yet Acceptable)—giving consumers the most advanced design possible while not overstepping into the unfamiliar. But Loewy's work sparked a movement that inspired other designers. Collectively, this new look became known as Streamline Moderne. It stripped the previously popular Art Deco style of its straight lines, angles, and ornamental trappings to focus on polished, curved shapes and long horizontal lines that took their cues from aerodynamics. And while Art Deco was seen as a style for a slender sliver of the upper class, Streamline Moderne was one for all, manifesting itself in everyday objects like radios, telephones, clocks, and other household items. For 1930s consumers in the grip of The Great Depression, streamlining screamed progress and the promise of a better future. That look still has its fans. Airstream aluminum traveling trailers, for example, trace their design to the 1930s and still have their admirers today.

Perhaps Loewy's most significant relationship was with the Pennsylvania Railroad, one that started small with the design of new trash cans but quickly grew to include locomotives. For Loewy, this was a childhood dream come true. Loewy rode in the cabs, holding a stick out the window with a white ribbon attached to measure airflow. In 1934, Loewy critiqued the Pennsylvania Railroad's initial design for its new GG1 locomotive as "ugly, disconnected, and full of rivets." Instead, Loewy recommended welding the locomotive shell onto the chassis to eliminate the unsightly rivets and added the "cat whiskers" speed lines to suggest motion. Loewy also included toilets for the engineers. The GG1 was a hit with passengers and proved durable as well, staying in service until 1983. The GG1 success prompted the Pennsylvania Railroad to invite Loewy to design its experimental S1 locomotive. The streamlined look of "the Big Engine" was thrilling even when it was standing still.

Not to be undone, the New York Central Railroad in 1938 debuted the 20th Century Limited, designed by Brooklyn-native Henry Dreyfuss, a talented designer in his own right (the Princess telephone and the Polaroid SX-70 Land camera are among his creations) and among the many designers who adopted the streamlined look. The phrase "red-carpet treatment" is derived from the crimson carpet passengers walked on to board the train. Railroad fans consider the Dreyfuss design to be the most beautiful train ever built. Both the S1 and the 20th Century Limited epitomized luxury with gorgeous interiors and became iconic symbols of the age, although the 20th Century Limited may be considered the more successful of the two as it stayed in service until 1967 while the S1 retired in 1945.

Loewy continued to dream big. For the futuristic-oriented 1939 World's Fair in New York, Loewy designed a rocket gun for the Chrysler exhibit that would shoot passengers to London in an hour in a streamlined module. As World War II began, Loewy was a design consultant to over one hundred companies and had a staff of three hundred, with offices in New York, London, Chicago, Sao Paulo, and South Bend, Indiana.

A decade later, Loewy's designs seemed to have permeated every aspect of American life. A 1949 cover story in *Time* magazine captured that deep influence by describing Loewy's day in detail, surrounded in his luxurious Manhattan apartment by products of his own design: the alarm clock, the Schick electric razor, the Lux soap wrapper, and a tailored grey suit with replaceable inch and a half cuffs.

A new 1950 convertible Studebaker Loewy designed brought him to work—his relationship with the car company began in 1936 as a design consultant. That car had special flamboyant touches like a plastic tailfin, recessed door handles, porthole windows, and a tiny, gold grilled air scoop above the hood ornament. Those touches were designed to get pedestrian tongues wagging. "The car must look fast whether in motion or stationary," said Loewy. "I want it to look as if it is leaping forward. If it looks 'stopped' it is a dead pigeon."

Loewy's affiliation with Studebaker continued until its demise in 1963. Loewy moved automobile parts that had once been outside the vehicle, like fenders, inside one sleek shell, anticipating modern automotive design trends. Loewy's aerodynamic designs of "personal" cars for Studebaker like the Starliner and later the Avanti models laid the groundwork for future cars like the Corvette and Mustang. Loewy also designed a distinctive domed bus for Greyhound in 1954 that gave passengers a more picturesque view. Loewy's other transportation design initiatives include multiple interiors for ships and for the Concorde supersonic plane.

Loewy's persona, right down to his mustache, was a key element of his selling strategy. Even his arrival at the office made an impression. Loewy "was a very nice man," recalled William S. Gillis, who worked for Loewy as an assistant stock room manager in 1950. "His wife Viola was a beautiful person many years his junior, but she loved him very much. His chauffeur was also the picture of perfection, dressed in boots and a peaked hat." Loewy paid his employees well and created a profit-sharing system. Many of Loewy's designers went on to establish their own firms.

Loewy had been married for about a year to his then twenty-eight-year-old wife when the *Time* story appeared. The article described Loewy as a suave but shy figure, fit, grey-haired, and medium-sized at five feet ten inches tall. Loewy spoke "in a subdued voice that is, at the same time, apologetic but compelling. His face is reposed, gentle, sad, and as inscrutable as a Monte Carlo croupier."

Loewy also knew how to present companies in the best light using graphic design. Loewy's work in graphic design spanned decades and his success was often due to simplifying a logo to its basics, a tactic that many in hindsight see as a precursor to modernism. Examples include graphic design that became iconic like the one for the Shell oil company that essentially remained unchanged for fifty years. Loewy also designed the Exxon logo that appeared in 1972. Loewy did graphic design work for Canada Dry and Coca-Cola soft drinks, the United States Postal Service, TWA, Nabisco, and others, putting these company logos into the public's consciousness. Among the most recognizable of Loewy's graphic brandings is the distinctive blue and white livery of Air Force One created for the Kennedy administration.

Loewy moved into the space age by designing interiors for NASA's Apollo missions and the Skylab space station, the latter notable for his insistence on private quarters for crew members. Loewy worked relentlessly until he sold the company in 1976 and in retirement, divided his time amongst homes in Monte Carlo and Palm Springs, where his indoor/outdoor pool design dazzled Californians.

"America's designer," as Loewy is often called, died in Monte Carlo at the age at ninety-two. Loewy's impact on American life was monumental. Loewy's MAYA design philosophy echoed into the future. One example is how Apple's gradual evolution of its iPod design prepared people for the iPhone. There's no better summary of Loewy's life than the title of his autobiography: *Never Leave Well Enough Alone.*

14

Scrabble Games the World (1938)

Figure 14. *Scrabble* inventor Alfred Butts and promoter James Brunot playing on a large *Scrabble* board. *Source*: LOOK Magazine Photo Collection, Library of Congress, Prints and Photographs Division, LC-DIG-ppmsca-68991.

If you recognize *obia* as a vowel dump, know *rhythm* can't be beat because it has none, that a *fez* is really cool, that the real value of a word like *qindar* is that it has no "u," and the mother of all words is *oxyphenbutazone* played across the top of the board for 1778 points with bonuses, then you're a fan of a game with letters on tiles called *Scrabble*. If you're not, then you're in a minority. It's estimated that three out of five American households own the board game. It's a favorite game of former presidents: Bill Clinton and Barack Obama are fans as was Richard Nixon. About thirty thousand people per hour start a game. Some 150 million *Scrabble* games and counting have been sold worldwide (in twenty-nine languages) since its inception in New York. And about one million tiles are lost somewhere.

But in the beginning, *Scrabble* was a game no one wanted to play.

Simply put, *Scrabble* is a word game where players score points by placing letter tiles, each inscribed with a numerical value, on a game board to form words found in the dictionary. *Scrabble* is played on a fifteen-by-fifteen grid of squares with varyingly colored premium squares multiplying the number of points awarded. The game flows like a crossword puzzle and plays left to right or downward. Players also can use an opponent's letters to form words of their own. Two blank tiles can be used to substitute for any letter. The game is over when all the letters have been drawn and one player uses his last letter or when all possible plays have been made.

Scrabble's inventor, Alfred Mosher Butts, was born in Poughkeepsie, New York, in 1899. In 1932, Butts was down on his luck like many other people suffering through the Great Depression. Butts was a man with a lot of time on his hands. The Depression had stalled his career as an architect, and he was subsisting as a part-time statistician. Butts did a lot of reading in his fifth-floor walkup in Jackson Heights, Queens, and, according to the New England Historical Society, it was a passage in an 1843 short story called *The Gold Bug* by the esteemed writer Edgar Allan Poe that gave Butts an idea. A character in the story decodes a message by comparing symbols to letters: "Now, in English, the letter that most frequently occurs is *e*. Afterwards, the succession runs thus: a o i d h n r s t u y c f g l m w b k p q x z," wrote Poe. Poe omitted the letters *v* and *j*, from the cypher pirate Captain William Kidd (circa 1645–1701) allegedly penned in *The Gold Bug*.

Butts realized that therein he had the makings of a game that would combine chance and skill and might prove as popular as card games and crossword puzzles. But for Butts to pull this off he needed to do research to assign a fair numerical value to any given letter and to determine how frequently any given letter should appear in the game. Butts figured out the right balance of letters by analyzing letters used in two newspapers, the *New York Times* and the *New York Herald Tribune,* and a magazine called the *Saturday Evening Post*. Butts sampled 12,082 letters and 2,412 words to come up with a statistically valid breakdown. It took Butts years to complete the analysis.

Then Butts made two mistakes. He called the game *Lexico*. And there was no board.

For four years, Butts tried to sell *Lexico* to big game makers like Parker Brothers and Milton Bradley. He drew blanks. A few friends bought the game. Butts wised up and added a board so words could be created crossword-puzzle style. Butts made the game himself, hand-lettering each tile and gluing them to pieces of balsam wood. Butts had calculated the number of times a letter should appear: twelve tiles for the letter *E* as the most used, one tile each for least-used letters, and varying amounts in between for the others. There are only four *S* tiles so the game wasn't too easy. Numerical values were in inverse proportion to their frequency, which is why one *Q* is worth ten points and the twelve *E* tiles are only worth one point each. Butts hand-crafted small racks to hold the letters. The letters *v* and *j*, absent from *The Gold Bug* story, were valued at two points and one point respectively.

Still, Butts hadn't quite hit the mark. Butts called the game *Criss-Cross* and sold each one for two dollars. Butts played the game with his wife Nina who turned out to be the better player.

Butts didn't sell many games of *Criss-Cross*, but one buyer turned the game around. James Brunot was looking for a small business to keep him occupied after retiring as a social worker in New York and Washington where he had served as executive director of the president's War Refugee Board in the 1940s. Brunot and his wife Helen were fans of *Criss-Cross* and had been avid players over the years.

Brunot contacted Butts, who by this time had resurrected his career as a New York architect with Holden, McLaughlin & Associates. Among his designs were a housing project on Staten Island and a library in Stanfordville, New York. When Brunot showed up looking to mass produce the game, Butts was more than happy to oblige in exchange for patent rights and a share of the royalties.

Brunot, as a devoted player, had an idea of how to tweak the game. Brunot came up with the now iconic color scheme of pastel pink, baby blue, indigo, and bright red. Brunot moved the start point to the center of the board instead of the upper-left corner and added a fifty-point bonus to the rules for using all seven tiles on a player's rack to make a word. Then Brunot made his best decision: he changed the name of the game to *Scrabble*. The Cambridge University dictionary defines the word *scrabble* as the act of using your fingers to find something that you cannot see. The perfect name as it turned out. Brunot copyrighted and trademarked it in 1948.

Success seemed assured but it wasn't. The Brunots initially made just eighteen games per day working out of their living room of their house in Newtown, Connecticut, but the business soon outgrew their home. "Couldn't move," Brunot told an interviewer in 1953. "No room for anything but boxes and racks and tiles."

Brunot moved operations to an abandoned schoolhouse nearby where production increased to twelve games per hour. Friends helped them stamp letters on the wooden tiles. In 1949, Brunot made 2,400 games but lost $450.

Still the game began to attract some hard-core fans and sales slowly increased. Then fortune struck. In the summer of 1952, Jack Straus, president of Macy's, the giant New York City department store, was introduced to the game by friends while on vacation. When Straus returned to work, he asked if Macy's sold the game. It didn't. Straus placed an order for two thousand.

The Macy's order was like an explosion. *Scrabble* quickly became the toast of the gaming world. Sales went from 4,853 in 1951 to almost four million by 1954. *Scrabble* was made for foreign languages and there was a braille edition. Brunot quickly added staff with thirty-five employees working two shifts. But at its best, Brunot's crew could only produce six thousand games per week. Customers were put on a waiting list.

Scrabble clearly had grown beyond a leisurely business run by someone in semi-retirement. A big-time player needed to take over.

Brunot licensed *Scrabble* to a games manufacturer called Selchow and Righter, which ironically had turned down the game when Butts had pitched to them many years earlier. Selchow and Righter produced *Scrabble* for the next thirty years. Brunot cashed out in 1971 and Butts sold his patent rights as well. They were both millionaires. Butts, ever the statistician, figured he earned about three cents per set. "One third went to taxes," Butts said. "I gave one third away and the other third enabled me to have an enjoyable life." Butts tried his hand at another game in 1985 simply called *Alfred's Other Game* but it failed commercially. Butts became an amateur artist, drawing New York scenes that were then reprinted on architect's linen using a blueprint machine. Six of them were acquired by the Metropolitan Museum of Art. Butts passed away in 1993 a few days short of his ninety-fourth birthday. Brunot died in 1984 at the age of eighty-three.

In 1986, the rights to *Scrabble* were sold to Coleco Industries but the company went bankrupt three years later. *Scrabble* was then purchased by Hasbro, owner of Milton Bradley, the leading games maker in the United States. Mattel wound up with international rights under a separate deal. *Scrabble* continued it popularity unabated. Among the variations developed was a game board with revolving table, a travel-sized edition, a junior version for children, and even a television show. Hasbro and Mattel sponsored major tournaments—scientists have studied ranked players as a way of understanding how language is processed in the brain. (Regrettably, a research team at the University of Calgary concluded that being good at *Scrabble* doesn't necessarily make you good at anything else.)

Scrabble, of course, can now be played online (although its appearance as an app was greeted with mixed reviews) and there is a website. There is even a dedicated dictionary for words allowed to be used—a reference tool that now excludes slurs. Remarkably, *Scrabble* has otherwise remained essentially unchanged in the decades since it adopted its current moniker. For some, *Scrabble* holds a deeper meaning than just a board game. "You must operate on the board in front of you with the constraints that you have," writes the *Guardian*'s Richard Godwin, a self-confessed former *Scrabble* addict. "As in life." For most of us, though, it's just fun. By the way, the word *oxyphenbutazone*, an anti-inflammatory drug, is the game's holy grail (theoretically possible but never used in a game).

15

The Invisible Woman Makes
Gone with the Wind a Hit Movie (1939)

Figure 15. Katharine Blodgett demonstrating her invention, 1938. *Source*: Smithsonian Institution Archives, Accession 90-105, Science Service Records, Image No. SIA2007-0282.

"After all, tomorrow is another day." When actress Vivien Leigh in the role of Scarlett O'Hara tearfully utters those last words in the classic 1939 film *Gone with the Wind* (*GWTW*), few knew that another woman had a large role behind the scenes in the film's success. Audiences applauded the crystal-clear cinematography produced by the "invisible glass" developed by pioneering scientist Katharine Burr Blodgett. Her invention revolutionized the movie business and even gave the military an edge in later conflicts. *GWTW*, while controversial in some of its on-screen depictions, remains the highest-grossing film of all time, adjusted for inflation.

While Katharine Blodgett brought a newfound clarity to the world, her own life began under a dark cloud. Blodgett's father, a successful patent lawyer at General Electric (GE), was murdered during a burglary in the family's Front Street home in Schenectady, New York, shortly before her birth on January 10, 1898. GE offered a five-thousand-dollar reward for the arrest and conviction of the killer, but the suspect hanged himself in his jail cell pre-trial in Salem, New York. Blodgett's financially secure mother moved to France with Katharine and her older brother George Jr. in tow. Eventually, the family returned to New York City where Katharine finished her early education.

The major turning point in Blodgett's early life came in 1917 when a former colleague of her father's and future Nobel Prize laureate Irving Langmuir gave her a tour of GE's research laboratories. Blodgett was hopeful that manpower shortages caused by World War I might open opportunities for women in research. Science, Blodgett believed, offered the best economic opportunities for women and she arrived with a physics degree in hand from Bryn Mawr, determined to avoid a high school or women's college teaching career most other women with a scientific bent were channeled toward. At the time, industry was reluctant to invest in training high-level women as it was feared they might marry and quit to raise a family.

Langmuir, a native-born Brooklynite who had been with GE since 1909, promised Blodgett a research position if she completed her higher education. Langmuir proved to be a genius spotter. Blodgett obtained her master's degree at the University of Chicago where she worked on the development of gas masks, which saved many lives during World War I. In 1918, Blodgett became the first female scientist to be hired at GE's research lab in Schenectady. Langmuir had begun work on a new field of research in chemistry that focused on the development of uniform oily films just one millimeter thick. Blodgett worked as Langmuir's assistant, but he soon encouraged her to further her own studies. Langmuir had a university in mind. In 1926, Blodgett became the first woman to receive a PhD in physics from the University of Cambridge in England where she rubbed science elbows with the likes of Nobel Prize winner Sir Ernest Rutherford. Her dissertation was on the behavior of electrons in ionized mercury vapor.

Blodgett returned to GE where she worked with Langmuir on improvements to the light bulb. The pair's studies of electrical discharges in gases helped lay the foundation for particle physics. Blodgett quickly settled into Schenectady life. In her free time, Blodgett acted in plays with the Schenectady Civic Players. She became an active conservationist, an avid amateur astronomer, and passionate gardener with a summer home on Lake George. And as a member of the Zonta Club, founded in Buffalo in 1919, Blodgett aided the careers of other professional women. Katharine Blodgett Gebbie, namesake and herself a gifted scientist who had a stellar career at the National Institute of Standards and Technology, recalled her aunt inspired her interest in science as a child as she "always arrived with suitcases full of apparatus with which she showed us such wonders as to how to make colors by dipping glass rods into thin films of oil floating on water."

In 1932, Langmuir won the Nobel Prize for his research into surface films. Blodgett joined Langmuir, who called her a "gifted experimenter," to further develop the research. As a warm-up, Blodgett invented a gauge that could accurately measure a film's thickness by its color. Blodgett's most important invention was within sight. Blodgett developed a technique to transfer the soap film from a water surface to a solid surface such as metal or glass. By repeating the process, Blodgett was able to build up films of barium stearate layer by layer, molecule by molecule, up to about three thousand layers.

Langmuir-Blodgett films, as they became known, changed lenses and Hollywood in particular forever. At that time, normal glass reflected a lot of light. Blodgett discovered that if she added forty-four layers of soap film to the surface of glass, with each layer equal to the thickness of one molecule, it would become ninety-nine percent transparent. The film coating Blodgett developed was one-quarter the average wavelength of visible light. With the film coating in place, any light that reflected off the glass would travel half a wavelength farther than the light reflected off the film's surface. Bottom line: reflections were canceled out. Blodgett had invented non-reflecting glass.

Further engineering led to the development of a coating that would not wipe off. GE called it "invisible glass" and Blodgett was nicknamed "the invisible woman." When the coating was applied to camera lenses and projectors, the improvement in picture quality was revolutionary. *Gone with the Wind* was the first color film to be projected through a lens based on Blodgett's invention. In 1940, Blodgett received a patent for "Film Structure and Method of Preparation," just one of seven she would receive during her career. The patent title belies its impact.

Invisible glass quickly became indispensable. Within a few years, the entire movie business was using invisible glass for cameras and projectors. With World War II in motion, submarines periscopes and reconnaissance spy planes relied on it. Blodgett's other wartime contributions included the creation of a better smoke screen that used just two quarts of oil to cover several acres and a method of deicing airplane wings. Blodgett dabbled in weather science, creating a device to rapidly measure humidity as weather balloons quickly ascended.

Blodgett's enduring legacy is invisible glass. Variations on Blodgett's invention are used in cameras, eyeglasses, microscopes, windshields, telescopes, picture frames, and basically anything that requires a transparent surface. Invisible glass won Blodgett numerous awards and three honorary doctorates from Brown University and Elmira and Russell Sage colleges. In 1943, Blodgett was recognized as one of the one thousand most distinguished scientists in the United States. Blodgett's rare status as a female inventor at that time drew public interest. Profiles of Blodgett appeared in publications like *Time* and the *New York Times*. Even her hometown of Schenectady honored her with Katharine Blodgett Day on June 13, 1951, and in 2008, a local school opened in her name. In

1972, the Photographic Society of America presented her with its annual achievement award. In 2007, Blodgett was inducted into the National Inventors Hall of Fame.

Blodgett continued to work with Langmuir until his death in 1957 and she was very close with his family. Blodgett herself never married. As chronicled by Harvard University's Radcliffe Institute, Blodgett lived in a "Boston marriage" with Gertrude Brown, who came from an old Schenectady family and later Elsie Errington, an English-born director of a local girls school. A Boston marriage was not necessarily a romantic relationship in those days and generally involved two wealthy women of independent means. "The household arrangement freed Blodgett from most domestic responsibilities—except for making her famous applesauce and popovers."

Blodgett retired from GE in 1963 and subsequently added quirky poetry to her interests. "That formaldehyde polyvinyl/If you eat it, you'll dine ill/One night at a party/When all the guests ate hearty/By actual count it made nine ill." The end credits rolled when she died on October 12, 1979, at the age of eighty-one.

16

Batman Is Born in the Bronx (1939)

Figure 16. *Batman* Issue no. 1. *Source*: United States National Archives and Records Administration.

It turned out to be a million-dollar idea. Actually, it was more like $2.2 million. That was the 2022 auction price of a near-perfect copy of *Batman* no. 1 comic book published in 1940 for ten cents. Less than fifty copies are known to exist.

Batman first appeared in *Detective Comics* no. 27 in 1939, a comic book with an estimated value of over one million, according to publisher DC Comics. (*Detective Comics* would give the publisher its name.) Since then, the caped crusader has appeared in numerous films that have raked in billions of dollars. Batman also has graced the small screen with a variety of television shows.

Not bad for a superhero born in the Bronx.

Yet while Batman fights against justice, the character's story is marred by injustice. One of Batman's creators turned out to be a villain. Much like Bruce Wayne was Batman's secret identity, the identity of the true creator of Batman remained a secret for decades.

The Batman saga actually begins with the success of Superman, introduced to the world in 1938. Superman, sporting a look inspired by circus strongmen and wearing a cape favored by fictional alien races, was essentially a benevolent super god that fired the imaginations of Depression-era readers. DC Comics, known then as National Allied Comics, was looking for similar heroes.

Much of the true story behind the creation of Batman is due to the dogged decade-long detective work of author Marc Tyler Nobleman and his book *Bill the Boy Wonder: The Secret Co-Creator of Batman* in 2012 and a documentary inspired by the book called *Batman And Bill* in 2017.

Bob Kane, a young animator, came up with the initial concept for Batman but he knew it was weak. Born in New York City in 1915 as Robert Kahn to parents of East European Jewish descent, Kane studied art at Cooper Union in New York City and had a job as an illustrator for the comic book packager Eisner & Iger.

Kane's Batman looked very different from the one known today. Kane's Batman was a blonde-haired man in red jumpsuit and a domino mask with two rigid bat wings attached to his back. Kane would later cite the legend of masked swashbuckler Zorro, the drawings of Leonardo da Vinci, and a 1930 film called *The Bat Whispers* as the source of his inspiration. The real inspiration for the Batman was a young writer whom Kane had met at an alumni party for DeWitt Clinton High School graduates in the Bronx.

Bill Finger was born in 1914 as Milton Finger in Denver, Colorado, but his family moved east to the Bronx. Like Kane, Finger had legally changed his name to disguise his Jewish heritage as anti-Semitism spiked in the years prior to World War II. They weren't alone. Finding employment in mainstream advertising and publishing proved to be difficult, many creative Jewish artists gravitated toward comic books, what was considered the runt of the literature litter. Jerry Siegel and Joe Shuster, the creators of Superman, were both Jewish, as were Jack Kirby and Joe Simon, the creators of Captain America (who first appeared in 1941). Stan Lee and Jack Kirby would create such superheroes as the Hulk, the Fantastic Four, X-Men, the Avengers, and many others during their careers. The notion of a secret identity—a man of action disguised as a man of inaction like Clark Kent or Bruce Wayne—was based on their own dual identities. In the late 1930s, the anti-Semitic broadcasts of Father Charles Coughlin had many listeners. In 1939, a pro-Nazi rally in Madison Square Garden attracted twenty thousand attendees. Batman, Superman, and Captain America would soon join the fight against the Nazis.

In 1939, Kane asked Finger, who already had a few gigs as a comics ghostwriter but also worked as a part-time shoe salesman, to look at his Batman drawings. Finger began to suggest refinements. Poe Park was a regular meeting place. Perhaps some dark inspirational shadow manifested itself as Poe Park is named after the famous author of the macabre, Edgar Allan Poe, who rented a cottage on the spot from 1846 to 1849, which still stands today. A clear vision of the Dark Knight soon emerged.

Finger suggested giving Batman a black-and-grey outfit with a bat symbol, a cape instead of wings, gloves and a bat-like cowl inspired by the look of the popular *The Phantom* newspaper comic strip hero. Like Superman, Batman would have an alter ego in millionaire Bruce Wayne, a name derived from a melding of Robert Bruce, the Scottish patriot, and Mad Anthony Wayne, a celebrated American general in the Revolutionary War. Finger was known for his meticulous research for the back stories of his creations. Batman was transformed from a super-vigilante envisioned by Kane into a super-detective protector of Gotham City, a nickname for New York City coined by author Washington Irving in 1807 (although Finger says he got the idea from spotting the name "Gotham Jewelers" in a phone book). The only carryover from Kane's initial idea was the yellow utility belt. But perhaps the real genius move was to make Batman a superhero devoid of superpowers, making Batman the opposite of the god-like Superman and more identifiable with readers.

DC Comics loved the Batman concept and put development on the fast track. Finger wrote the script for Batman's first appearance in *Detective Comics* no. 27 while Kane supplied the art. The Batman origin story is well-known: after witnessing his parents' death by a mugger outside a movie theater, young Bruce Wayne promises to rid the world of the evil that took their lives. What many called the real crime, however, was Kane's deal, in Finger's absence, with the publisher that stipulated in a contract that he would be given sole credit for Batman.

Finger would continue to develop the Batman mythos. Batman's sidekick Robin also was a Finger creation, debuting in *Detective Comics* no. 38 in April 1940. As Finger later recalled, Robin was an outgrowth of a conversation with Kane. Like Batman, Robin was an orphan. "The thing that bothered me was that Batman didn't have anyone to talk to and it got a little tiresome always having him thinking. I found that as Batman went along Batman needed a Watson to talk to. That's how Robin came to be." Robin was to Batman what Dr. Watson was to Sherlock Holmes as a sidekick and explainer of the detective work. Kane suggested putting a boy into the strip to attract younger readers. Credit for the name is claimed by Jerry Robinson, a new member to the Batman team who helped develop the Joker, the evil villain for the first issue. An unsung hero is the editor who changed the Joker's planned demise into a survivable knife wound, creating a legendary nemesis for Batman that would appear repeatedly in story lines.

Superman, Batman, and Robin would soon join forces, depicted with a famous cover of *World's Finest Comics* no. 9 depicting the trio hurling baseballs at the faces of Hitler, Mussolini, and Tojo, the leaders of the Axis alliance of Germany, Italy, and Japan. The adventures of Batman continued with Finger penning the scripts. Kane, meanwhile, had quickly farmed out the actual drawing of the character to ghost artists, a common practice at the time. Finger created or co-created many of the characters in the Batman universe, most notably Batgirl, the Penguin, and the Riddler as well as developing key elements like the design of the Batmobile, the Batcave, and Gotham City. Finger wasn't exclusive to the Batman story as he is credited with creating or co-creating the Superboy and Green Lantern superheroes while having a hand in the creation of many more like Wonder Woman. It's estimated that Finger wrote over one hundred fifty issues of *Batman* and perhaps as many as six hundred issues of comic books for DC Comics overall. He had a finger in a lot of pies.

Batman, however, was among the comic book heroes attacked as immoral in the 1950s as the United States was gripped by Red Scares and the witch hunts of McCarthyism. Batman and Robin being gay were among the charges levied by Fredric Wertham, a bespectacled German-American psychiatrist practicing in Harlem, in his 1954 book *Seduction of the Innocent*. The book created an uproar. Comics led to juvenile delinquency, screamed detractors. A Girl Scout troop in Missouri burned comic books en masse. A Senate subcommittee investigated the industry, focusing particularly on the links of comic book distributors to organized crime. Wertham testified that comic books scared him more than Hitler. Comics became front-page news and public opinion shifted negatively away from comic books. A Comics Code Authority was created to monitor comic-book content and comics required its seal of approval. Among the requirements were a ban on slang and vulgar language, sexy images, and stories that dealt with racial or religious prejudice, which effectively removed non-whites from the comic book universe. They joined werewolves, zombies, and vampires who also were banned. Many comic book publishers went out of business.

Television also was an enemy of comic books as a competitor for leisure time. The two mediums brokered a peace of sorts with the debut of *Batman* as a TV series in 1966 that was played for campy kicks. Finger, meanwhile, briefly shed his Clark Kent meekness in 1965 to tell an early comic-book convention of his key role in creating Batman. Kane hammered Finger in a published response, denying Finger's claim and insisting he was the sole creator of Batman. The credit to Kane by DC Comics backed him up. When *Batman* aired on TV, Kane raked in a fortune and continued to do so when Batman returned to the big screen with the widely acclaimed 1989 *Batman* that rebirthed the franchise in its dark glory.

Finger returned to obscurity, eking out a living as a screenwriter of films with titles like *The Green Slime* and *Track of the Moon Beast* as well as writing for TV series like *77 Sunset Strip* and *Hawaiian Eye*. Finger passed away virtually penniless in his Manhattan apartment in 1974. Finger's remains were claimed by his son Fred and were cremated. Finger's ashes were spread in the shape of bat on a beach in Oregon according to his wishes.

Kane eventually admitted in his 1989 autobiography *Batman & Me* that Finger deserved credit for creation of Batman. Kane died ten years later. But it wasn't until the publication of Nobleman's book—a feat of detective work worthy of Batman himself—that the full extent of Finger's key creative contributions became known to the general public. Interest in Batman was high as 2012 saw the debut *The Dark Knight Rises*, director Christopher Nolan's final entry in his Batman trilogy, a film series that drew inspiration from 1970s comic book artist Neal Adams's photorealistic style and from artist-writer Frank Miller's graphic novel depiction of Batman as a darker, grittier character. In Miller's and others' redrawing of the character, Batman re-assumes the more ruthless and psychologically complex attributes exhibited in 1938 that had been softened over the years. After a campaign led by Finger's granddaughter Athena, DC Entertainment in 2015 formally acknowledged Bill Finger's role in the creation of Batman and credited him in all subsequent Batman comics and films.

On a chilly Friday morning in December 2017, Bill Finger received a final long overdue accolade. In a ceremony attended by friends, fans, and family, the south corner of One Hundred Ninety-Second Street and the Grand Concourse in the Bronx was named Bill Finger Way. That's the entrance to Poe Park. The Batman story had come full circle. Batman's Gotham City, meanwhile, had become home to so many superheroes they practically needed their own neighborhood. The list of superheroes born or residing in New York City, largely an expression of the rival Manhattan-based Marvel Comics group, grew to include the Avengers, Spiderman, the Fantastic Four, Daredevil, Captain America, Jessica Jones, Doctor Strange, Luke Cage, Hulk, Iron Fist, Punisher, Spawn, Black Panther, Watchmen, Teenage Mutant Ninja Turtles, and Catwoman. For superheroes, there is no place like New York.

17

The Bloodmobile Drives to War (1940)

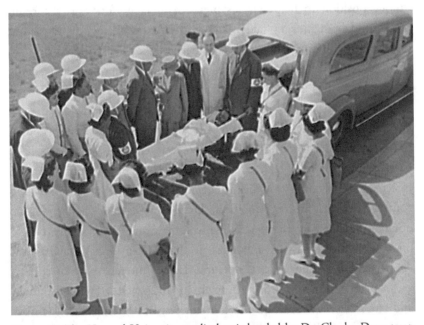

Figure 17. The Howard University medical unit headed by Dr. Charles Drew treating a patient from the Bloodmobile in a practice drill, 1943. *Source*: Library of Congress, Prints and Photographs Division, FSA/OWI Collection, LC-USE6-D-010067.

If you've ever donated or received blood, then you have a pioneering African-American doctor to thank. Dr. Charles Drew's development of a blood bank has saved countless lives since its creation at a New York hospital. Drew also is remembered for his stand against racism when health authorities of the period decided that blood donations must be separated according to the color of the donor's skin.

In the early days of World War II, Great Britain reeled under "the Blitz" as waves of Nazi bombers attacked the country. London was bombed for fifty-six out of fifty-seven days straight. Major industrial cities like Liverpool, Coventry, Birmingham, and Manchester were targeted along with smaller municipalities. Adolf Hitler believed this bombing campaign would force the United Kingdom to capitulate. They didn't but casualty figures were enormous. There was a dire need for blood for lifesaving transfusions. But there wasn't enough of it.

Across the "pond" in New York City, Dr. Charles Drew was in a unique position to help. Drew's presence in New York was the result of a roundabout journey taken largely to circumvent racism. Drew was born in Washington, DC, in 1904, the oldest of five children born to a father who worked as a carpet layer and a mother with a teaching degree. Drew was an exceptional athlete and attended Amherst College in Massachusetts on an athletic scholarship for both football and track and field, captaining the latter. As one of thirteen Blacks in a student body of six hundred, the racial climate was far from ideal, especially when facing opposing teams. Drew excelled academically but it was an infected football injury requiring hospitalization that brought him into contact with his future profession. Drew's interest in science was no doubt influenced by the death of his sister from tuberculosis brought on by complications with influenza in 1920. Otto Glazer, chair of the Amherst biology department, nurtured Drew's interest in science.

After graduation from Amherst in 1926, Drew taught biology and chemistry and coached the football team at historically Black private Morgan College in Maryland, saving up to enter medical school. Drew discovered that he was two English credits short of the medical school entrance requirements for Howard University and while Harvard beckoned, that university wanted him to wait another year before being admitted as they only accepted a few non-white students per year. Not willing to wait, Drew chose to enter McGill University's medical school in Quebec, a decision likely influenced by the Canadian school's reputation for being supportive of non-whites. Five years later, Drew graduated second in his class certified as both a medical doctor and as a surgeon. A one-year residency studying shock and fluid resuscitation at Montreal General Hospital followed.

Drew wanted to further his surgical training in the United States but because of his race, options were limited. Only two postgraduate training programs were open to African Americans and Drew chose Freedman Hospital in Washington, DC, to be close to home, a fortuitous decision as his father died in 1935, making Drew the family's primary breadwinner. Drew added to his workload by becoming an instructor of pathology at the Howard University College of Medicine while also continuing at Freedmen's Hospital where he was assistant surgeon by 1938. Drew was being groomed for the chief of surgery position, but additional training was required. Funding from the Rockefeller Foundation brought Drew to New York's Presbyterian Hospital to further his surgical training and research. Drew enrolled at Columbia University to pursue his doctorate in medical science.

Drew, with his background in shock and fluid resuscitation techniques, fell in with Dr. John Scudder, a blood transfusion specialist. Together, they set up an experimental blood bank to research all aspects of blood preservation and transfusion methods. Drew's doctoral research, published in 1940, focused on every aspect of blood storage. The main problem was how to keep blood from spoiling. "Banking" blood for use when needed was one of the biggest medical challenges of the time as blood loses its utility soon after it leaves the body. Hospitals kept "blood bank" lists of registered donors who could be called when needed.

Drew discovered that blood cells were not what determined blood type and that plasma, the liquid portion of blood, being devoid of cells, could be given to anyone in need. Drew realized that unlike whole blood, plasma, the fluid portion of the blood, could be stored without refrigeration or deterioration during transport. Drew devised a method by which plasma could be dried and then reconstituted with distilled water when required. Blood plasma, while lacking oxygen-carrying cells, has critical proteins and antibodies that help stabilize blood pressure and regulate clotting. Drew created a blueprint for the procedure to be done safely and packaged without contamination. Scudder described Drew's "Banked Blood" dissertation as a "masterpiece." Drew became the first African American to earn a medical doctorate from Columbia University.

Emergency medicine was transformed. It came just in time.

In 1940, England's need for medical supplies to treat the injured in what became known as the Battle of Britain was desperate. While the United States had yet to enter World War II, many in the medical profession saw the direction in which events were headed. A cooperative group of New York hospitals created a "Blood for Britain" campaign. New Yorkers stepped up. Within a few weeks, ten thousand donors had appeared at eight hospitals. The need for a full-time medical director to standardize collection, storage, and transportation procedures became clear. Drew was the obvious candidate.

Drew rescued what threatened to be a messy operation. "Since Drew, who is a recognized authority on the subject of blood preservation and blood substitutes, and at the same time, an excellent organizer, has been in charge our major troubles have vanished," wrote a medical colleague. By January of 1941, just over 14,500 people had donated 6,151 liters of plasma, which were shipped in quart bottles to Britain.

Savvy doctors knew that with war raging in Europe and China, the "Blood for Britain" program was just a warm-up for what was to come. With its national scope and local chapters, the American Red Cross was the ideal organization to spearhead a blood-collection drive across the country. Once again, Drew was the obvious choice, becoming director of the first American Red Cross blood bank at Presbyterian Hospital and assistant director of blood procurement for the National Research Council. Perhaps his major innovation was the creation of a "bloodmobile," a van large enough for blood collection and refrigerated storage. Drew also developed a way to dry and package plasma and kits that medics could use to reconstitute the plasma for transfusions. The program was going well. Thousands of soldiers' lives would be saved. Drew became known as "the father of the blood bank."

But with a blood drive now national in scope, the question of the racial identity of the donors arose even though the science of blood typing was well established. Race didn't matter. Blood was blood. Despite knowing this, the War Department issued a directive: "For reasons which are not biologically convincing, but which are commonly recognized as psychologically important in America, it is not deemed advisable to collect and mix Caucasian and Negro blood indiscriminately for later administration to members of the military forces."

The American Red Cross adopted the policy, at first prohibiting Black blood donations but then labeling plasma with its racial origin. In April 1941, Drew resigned from both his positions at the American Red Cross and the National Research Council.

Drew was not an activist by nature, and he was clearly reluctant to publicly criticize the government during wartime. His silence as to the circumstances of his departure left many bewildered, especially African Americans. "It seems strange that his country could find no further use for the citizen who had been of such viral expert assistance in the critical hour," wrote W. Montague Cobb of Howard University. "One hears that it was thought that a Negro would not be acceptable in a high place in a national program."

By 1944, perhaps with the war's end in sight, Drew could no longer contain himself. Drew wrote to the director of the Federal Labor Standards to address the issue.

"I think the Army made a grievous mistake, a stupid error in first issuing an order that blood for the Army should not be received from Negroes. It was bad mistake for three reasons: (1) No official department of the Federal Government should willfully humiliate its citizens; (2) There is no scientific basis for the order; (3) They need the blood."

Drew amplified his objections in public. At one awards ceremony, Drew told his audience: "It is fundamentally wrong for any great nation to willfully discriminate against such a large group of its people." Drew added: "One can say quite truthfully that on the battlefield, nobody is very interested in where the plasma comes from when they are hurt. It's unfortunate that such a worthwhile and scientific bit of work should have been hampered by such stupidity." The American Red Cross ended its segregation of blood in 1948.

Drew served as a consultant to the surgeon general on the state of surgical facilities in Europe after the war. But Drew's main focus was on what he felt was his true mission in life, the training of the next generation of Black doctors and surgeons. Drew became chief of surgery and chair of the department of surgery at Freedmen's Hospital and renewed his teaching with Howard University. Drew also lobbied against the exclusion of Black doctors from the American Medical Association and other specialty medical societies. "The breaching of these walls and the laying of this road has not been, and is not easy," noted Drew.

Drew's educational mission was cut short by his death in the early hours of April 30, 1950, in a car accident in Burlington, North Carolina, while en route to a conference. Drew received prompt and competent care from white physicians at a local hospital, but his injuries proved too severe. A persistent myth developed that Drew was denied admission to the white hospital or that he was denied a transfusion because of his race but those allegations have been debunked numerous times, particularly by the three Black colleagues traveling with him. Drew was just turning forty-seven years old when he died, leaving behind a wife and four children. Among Drew's many posthumous recognitions is his inclusion in a Great Americans stamp series issued by the United States Postal Service in 1981. In the end, Drew's innovations surmounted racial differences. Blood banks are now taken for granted by all.

18

The Atomic Woman (1945)

Figure 18. Chien-Shiung Wu, 1958. *Source*: Smithsonian Institution Archives, Acc. 90–105, Science Service Records, Image No. SIA2010-1509.

She was called the Dragon Lady by her students at New York's Columbia University. Her colleagues knew her as a five-foot-tall powerhouse and called her the First Lady of Physics, the Queen of Nuclear Research, and the Chinese Marie Curie. Chien-Shiung Wu played a crucial role in the development of the atomic bomb and discovered that nature knows right from left, thereby changing science's view of the universe. And as a woman in a field largely dominated by men and who was often the first female to hold various positions, Wu was an ardent supporter of women in science. Wu herself would experience sexism firsthand as she was overlooked for the Nobel Prize.

"There is only one thing worse than coming home to a sink full of dirty dishes," said Wu. "And that is not going to the lab at all."

While Wu's major accomplishments occurred while she was a resident of New York City, Wu was born on May 31, 1912, in a small village near Shanghai in China. Wu began her education in a school run by her father, who believed in education for girls, unusual for the time and place, and with whom she was very close. Her father had decamped to the remote countryside after participating in China's failed Second Revolution. Wu first studied mathematics but switched to physics at university in China before becoming a researcher at the Institute of Physics of the Academia Sinica.

Encouraged by a mentor, Wu decided to study physics at a higher level than was possible in China at the time. Wu also was a student leader involved with protests against Japan's invasion of China beginning in the early 1930s so politics may have been a consideration as well. With financing from an uncle, the twenty-four-year-old Wu sailed from China to San Francisco in 1936. By the following year, nearby Shanghai was a battlefield. When Wu said goodbye to her parents, little did she realize that world events to come to mean she would never see them alive again. She wouldn't return to China for another twenty-three years.

Wu had initially planned to continue her studies in Michigan but a visit to the University of California, Berkeley changed everything. With an already impressive resume, Wu was offered the chance to study under Ernest O. Lawrence, one of the biggest names in physics. Lawrence would win the Nobel Prize for Physics in 1939 for his invention of the cyclotron particle accelerator and, along with fellow physicist Edward Teller, would found Lawrence Livermore National Laboratory in 1952. Wu would earn her PhD in physics in 1940 with a focus on cutting-edge nuclear technology. In 1942, Wu married a fellow Berkeley physicist Luke Chia-Liu Yuan, the grandson of Yuan Shikai, the first president of the Republic of China and self-proclaimed emperor of China. War in the Pacific prevented either of their parents from attending the ceremony.

During World War II, anti-Asian fervor rose to a fever pitch on the West Coast as the United States battled Japan in the Pacific. Wu and her husband decided to move to the East Coast where the employment atmosphere was more favorable. Yuan immediately was hired by Princeton University in New Jersey where he worked on radar development. Wu, after a brief stint teaching at Smith College, also joined the Princeton faculty, attracted by its world-class research facilities. Wu was the first female instructor at what was then an all-male school.

Wu's time at Princeton was cut short when she was invited to join the Manhattan Project as a senior scientist. Two Columbia University physicists reportedly spent a day interviewing Wu while trying not to be too specific about the actual job. Wu eventually brought them up short. "I'm sorry but if you wanted me not to know what you're doing, you should have cleaned the blackboards." She was immediately hired.

The Manhattan Project was the top-secret operation to build the first atomic bomb, an effort that involved thousands of scientists and engineers. The fear was that Adolf Hitler had been working on an atomic bomb since the 1930s and would eventually use it. While often linked to sites like Los Alamos, New Mexico, where the first atom bombs were built and tested, the Manhattan Project got its name because most of the early work was done in New York City. According to atomic age historian Robert S. Norris, there were ten sites employing five thousand people in Manhattan alone that, unbeknownst to city residents, were key locations when the project began. These included warehouses that stored uranium, laboratories attempting to split the atom, and the project's first headquarters across the street from City Hall. Manhattan was already home to a cadre of top European scientists who fled the Nazis and the city's piers were ideally situated to receive imported uranium from places like the Belgian Congo. Other nearby locations also were utilized—for example, a critical cyclotron was built at the Nevis Laboratories of

Columbia University in Irvington, New York, an easy train ride north from the city. Eventually, wise heads realized that nuclear research was better moved to more remote, less populated areas.

A lot of intellectual brain power remained in New York; however, including Wu who settled into a post at Columbia University where her work included improvements to radiation detectors. Wu is the only known Chinese person to have participated in the Manhattan Project. Wu's nuclear expertise was instrumental in solving a mystery at the Hanover Site in the southeastern steppes of Washington State. The B reactor worryingly began shutting down of its own accord. At the invitation of Enrico Fermi, creator of the first nuclear reactor, Wu helped determine that xenon-135, a product of fission, was the culprit so better control measures could be taken.

Wu's critical contribution, however, was the development of a process that separated uranium-235 and uranium-238 isotopes using a methodology called gaseous diffusion. Wu's approach was implemented on a gigantic scale at the K-25 site near Oak Ridge, Tennessee. This process is what turned a simple bomb into an atomic bomb. The Manhattan Project culminated in the detonation of atomic bombs over Hiroshima and Nagasaki, Japan, on August 6 and 9, 1945, ending World War II.

After the war, Wu remained at Columbia where her reputation grew as the best experimental physicist of the time with particular expertise in an aspect of nuclear physics called beta decay. New Yorker J. Robert Oppenheimer, the "father of the atomic bomb" who oversaw the Los Alamos production facility, called Wu "the authority" on the subject. In 1952, Wu became the first woman to have a tenured faculty position in Columbia's physics department. Much of her work involved proving or disproving other scientists' theoretical ideas. Her most famous work in this regard disproved the notion of parity—that Nature at its most basic level had no left- or right-handed bias and there is a fundamental symmetry in how Nature behaves.

Parity was a widely accepted notion across science. The trouble was that a pair of tiny particles called kaons were behaving badly. Tau and Theta were originally believed to be separate particles because they decayed into fundamentally different things. But as more data was examined, it became clear that Tau and Theta were actually one and the same. To have one particle produce offspring of different parities was akin to a dog giving birth to both puppies and kittens. Scientists flipped.

There are four forces that govern the universe. Three of them—gravity, electromagnetism, and a "strong" force hold things together. The fourth is a "weak" force that makes matter fall apart. In 1956, two theoretical physicists, Tsung Dao Lee of Columbia University and Chen Ning Yang of Princeton University, proposed the parity rule wasn't at work in nuclear reactions involving the weak force. The pair suggested an experiment to prove their hypothesis. The experiment was tricky and required someone with meticulous skills. They turned to Wu to make it happen.

Wu realized that to prove the theory she would need the use of a cryogenics laboratory and found one near Washington, DC, at the National Bureau of Standards and Technology that was suitable. Wu had a remarkable ability to keep the overall goal in mind while also attending to minute details. Wu designed the machines used for the experiment, wrote the laboratory procedures, and created the guidelines to measure the results. Wu and her team placed radioactive cobalt 60, a material Wu knew decayed by beta particle emission, into a strong magnetic field that would cause the spin axes of the atomic nuclei to all line up in the same direction. The cobalt was supercooled to minimize any random thermal motion. Tens of thousands of electrons were ejected during the course of the experiment. Most went in one direction. Nature knew its right from its left.

Scientists were gobsmacked, but Wu's results were quickly confirmed by other experiments at Columbia University and elsewhere. Wu's experiment had "changed the accepted view of the universe," she later said. "I had disproved one of the widely accepted 'laws' of the universe, the conservation of parity, by proving that identical nuclear particles do not always act alike." The experiment still bears her name.

Lee and Yang won the Nobel Prize in Physics the following year. Wu's critical work was not acknowledged. Nobel apologists note that theoretical scientists have often been favored over experimental practitioners like Wu even though Lee and Yang's theory would not have been proven without her. Most saw the omission as a gender-based snub, as did Wu herself. "Why didn't we encourage more women to go into science?" Wu publicly remarked at a 1964 symposium. "Unfriendly social atmosphere and psychological hindrance are at the root of the problem. I wonder whether the tiny atoms or nuclei, or the mathematical symbols, or the DNA molecules have any preference for either masculine or feminine treatment."

The Nobel Prize snub caught a lot of people's attention. The author and playwright Clare Boothe Luce echoed the opinion of many: "When Dr. Wu knocked out that principle of parity, she established the principle of parity between men and women." Luce was the wife of Time Inc. legend Henry Luce, and she would serve her country in Congress and as an ambassador to Italy. For most of her career, Wu's salary did not match that of her male colleagues. A book Wu wrote in 1965 called *Beta Decay* is still a standard reference for physicists.

Wu and her husband had a son named Vincent in 1947 who grew up to become a physicist at Los Alamos. Wu became a naturalized American in 1954. World War II and a civil war in China afterward dashed plans to visit China to see family. Later, her father told her not to return to communist China. Wu wasn't able to visit China until 1973, whereupon she discovered that the graves of her parents were destroyed and that her uncle and brother were victims of the Chinese Cultural Revolution. Wu was very respectful of her Chinese origins and wore Chinese-style qipao dresses most of her life, even making her own during the years when China was cut off from the outside world. Wu's New York apartment was home to jade statues and scroll paintings.

There is no doubt that the lab was Wu's favorite place to be, and she was known to sleep there on occasion. Her granddaughter Jada Yuan recounted in a 2021 memoir in the *Washington Post* a story that illustrates the case. Wu's students hoped to get the Dragon Lady (a nickname borrowed from a popular comic) out of the lab for even a short time. The students bought circus tickets for Wu and her son but Wu returned quickly, having given her ticket to her son's nanny.

While snubbed by Nobel, Wu became widely recognized for her groundbreaking work, adding research on sickle cell anemia to her list of interests. Wu became the first woman president of the American Physical Society. In retirement, Wu worked to get more women interested in science. Among her numerous honors are a National Medal of Science and a stamp issued by the Unites States Post Office with her image. In China, Wu was treated like a rock star when she visited. Wu died of a stroke at her New York City home in 1997 and was buried in her Chinese homeland.

19

The LP Makes Vinyl a Hit Record (1948)

Figure 19. Labels for Columbia Long-Playing Records. *Source*: Box 121, folder 3, Columbia Records Paperwork Collection, Motion Picture, Broadcasting and Recorded Sound Division, Library of Congress.

The man who set the long-playing (LP) in motion was named Peter Goldmark. And despite predictions of its demise, the LP vinyl record plays on.

While the LP record may be Goldmark's most enduring legacy, Goldmark is often described as akin to one of those "lone wolf" inventors from the nineteenth century with a work ethic that put him at his desk at 5:00 a.m. Goldmark was precise, creative, and prolific. A self-described maverick, Goldmark's career not only revolutionized audio but also impacted the development of television and the early days of the space age.

Goldmark was born in 1906 in Budapest and grew up in the turbulent, waning days of the Austrian-Hungarian empire that had been ruled by the Hapsburg monarchy for centuries. While little is known of his early years, Goldmark was raised in a classical music environment. His mother was a violinist and Peter learned the cello to become part of the household chamber music ensemble. Peter Goldmark's father Sandor was a businessman. Other relatives had already made their mark. Great-uncle Joseph Goldmark discovered red phosphorus used in making matches and invented the percussion caps for rifles first used in the United States Civil War. Another-great uncle, Karl Goldmark, was one of Hungary's greatest composers. The mix of science and music was in the blood.

Young Peter inherited an iron will from his mother Emma. With the breakup of the empire after World War I, Hungary seesawed between governments controlled by communists, monarchists, and right-wing nationalists. A "white terror" that targeted perceived anti-Hungarian elements like intellectuals and Jews—the Goldmarks were both—resulted in the execution of an estimated five thousand people and the imprisonment of another seventy-five thousand. During the fighting, rebels cruising the nearby Danube River repeatedly fired bullets through the open windows of their home as a warning to shut them. A string quartet was performing at the time and Peter's mother Emma refused to heed their warning until the quartet finished playing.

The Goldmarks fled the chaos to nearby Vienna in the newly-minted republic of Austria. Goldmark developed an intense interest in radio and continued his studies at a technical institute in Berlin, Germany and then received his PhD in physics in Vienna by 1931. Goldmark was already entranced by the promise of television. With a kit sold by a British inventor, Goldmark constructed a television receiver for experimental BBC broadcasts transmitted late at night. "The picture came through in postage-stamp size," Goldmark later recalled. "You could hardly make it out, it flickered so. It was also in color, all red. But it was the most exciting thing in my life."

Goldmark then made his way to London where he was turned down for a job by television pioneer John Logie Baird. Undeterred, Goldmark found a post as a television engineer at a rival radio company. Unfortunately, that company fell victim to the financial calamities of the Great Depression and Goldmark soon found himself back in Vienna.

Goldmark then had chance encounter with a CBS (Columbia Broadcasting System) correspondent who suggested Goldmark decamp to the United States. Goldmark arrived in New York City in 1933. Goldmark applied for a job at RCA (Radio Corporation of America) but was turned down for a post by legendary TV pioneer David Sarnoff. "This proved to be David Sarnoff's biggest mistake," Goldmark gleefully asserted later with a lack of self-effacement that some ascribed to his Hungarian heritage.

While it took Goldmark a few years to find his footing, by 1936 the new immigrant was chief television engineer at CBS (he became a naturalized citizen in 1937) and worked to refine the company's existing black-and-white television technology while also pioneering early work into color television. In the 1930s, television was still an experimental medium and radio was still king. Inspired by the lush color images of the film *Gone with the Wind*, a color TV system was developed by CBS under Goldmark's direction. The first color television broadcast (on August 7, 1940) was made from a CBS transmitter in New York.

World War II, however, forced Goldmark to redirect his energies. Goldmark became the technical director and then acting director of a laboratory in England that worked to develop electronic countermeasures and reconnaissance against the Nazis.

After the war, Goldmark returned to his work in color television. But while the color TV Goldmark developed was well-received from a technical standpoint, its incompatibility with an increasingly widespread network of black-and-white televisions already in people's homes doomed its adoption. An RCA system that was compatible won out but Goldmark's system, which won accolades for the quality of its image, would become the basis for the TV equipment used by Apollo astronauts in the 1970s to beam pictures back from the moon.

But almost like a soundtrack to his own life, Goldmark's interest in music never wavered and seemed to run in parallel with his fascination with television. Inspiration struck in 1945. "I was at a party listening to Brahms being played by the great Horowitz," recalled Goldmark in a 1973 autobiography. "Suddenly there was click. Somebody rushed to change records. The mood was broken. I knew right there and then I had to stop that sort of thing."

The mood-killing problem was that the 78 rpm discs of the day were very limited in their playing time—just five minutes per side. The 78 rpm records also were made using shellac, sourced from tree-dwelling insects, which while praised for its moisture-resisting properties proved to be very fragile. One drop would shatter them into pieces. Despite its drawbacks, 78 rpm records were the standard for decades.

That all changed in 1948. Leading a team of CBS engineers, Goldmark developed a long-playing (LP) record that was what a later generation might have called groovy. The LP record had grooves just 0.003 of an inch thick compared to the 0.01-inch thickness of 78 rpm records. The LP also played at a different speed: 33 1/3 rpm.

The LP, recorded on vinyl, was transformational. On June 21, 1947, at a press conference at New York's legendary Waldorf Astoria Hotel, CBS Board Chairman Ted Wallerstein unveiled a 12-inch, long-playing, unbreakable vinyl record with 22½ minutes of recording time per side. Not only did the new LP sound better than its predecessor, but it was also less expensive. One LP would cost $4.85 versus $7.25 for a collection of five 78s containing the same symphony. The first demonstration

LP featured a CBS secretary playing piano, an engineer on violin, and Goldmark on cello. The first LP for sale was available a week later—Mendelssohn's *Violin Concerto in E Minor* performed by the New York Philharmonic. But it was the subsequent recording of more popular fare like the 1949 musical *South Pacific* that brought widespread acclaim. By 1972, LP sales represented one-third of CBS revenue. Also in the CBS revenue mix were the seven-minute 45 rpm "singles" developed by RCA Victor in 1949. Phonographs or turntables that could toggle between the two speeds became a mainstay for hi-fi fans.

The LP is perhaps Goldmark's most significant invention, but his subsequent career was one of mixed results. A "highway hi-fi" record system developed for Chrysler cars couldn't overcome vibration challenges but laid the groundwork for car stereo systems of the future. Likewise, an early version of video recording laid the groundwork for the video cassette recorder to come.

Goldmark's main contribution as an inventor may have been as a prod to innovation, a role he acknowledged in his autobiography. "As I look back, I think my contributions were, somewhat ironically, not so much the invention itself or in innovation (a word I prefer because it means putting an invention to work) but in its gadfly impact on industry," he wrote. "The development of the long-playing record impelled the recording industry including RCA, the giant of the communications business, to change for the better its historical pattern of record production. My work in color television resulted, I think, in bringing color to the public a decade faster than it might otherwise have come, although not in the form intended. Finally, electronic video recording, though it ended up without the auspices of CBS, fired up the video cassette business into the potential multi-million-dollar business we are beginning to enjoy today."

Peter Goldmark died in an automobile accident, while driving alone, on the Hutchison River Parkway in Westchester, New York in 1977 at the age of seventy-one—just two weeks before his death he had been awarded the National Medal of Science from President Jimmy Carter. The obituary for the LP record was written in 1982 when digital compact discs became widely available. Since then, digital recording has morphed into various forms ranging from handheld devices to streaming services. The death of vinyl notices may have been premature however, as LP record sales in recent years have enjoyed a comeback with fans trumpeting a perceived warmer analog sound, a more tactile listening experience, and perhaps a more transitory human interaction as compared to eternal digital perfection. That human aspect would be a trait Goldmark would recognize. In a 1975 acceptance speech of a National Trustees Award at the Emmys television ceremony, Goldmark noted: "We still have to perfect that most important system of communications—from man to man."

20

How to Mend a Broken Heart (1960)

Figure 20. Wilson Greatbatch holding the original pacemaker. *Source*: Courtesy University Archives, University at Buffalo, State University of New York.

A nyone who claims New Yorkers are heartless doesn't know the story of the two New Yorkers who mended the hearts of millions. One, Dr. Wilson Greatbatch, seized upon a mistake to invent a working implantable pacemaker. At about the same time,

Dr. Robert H. Goetz performed the first coronary bypass surgery, an operation that he was only allowed to perform once and that would go unheralded for the next forty years. What they have in common besides their heartfelt focus is that they both first practiced their procedures on dogs, cementing the canine's reputation as man's best friend.

In 1958, Greatbatch was tinkering around in his backyard barn in Clarence, New York, just northeast of Buffalo. Greatbatch, an electrical engineer teaching at the University of Buffalo, was working on a piece of gear he hoped would be able to record heart rhythms. Greatbatch reached into a box of parts for a resistor to complete the circuitry. Accidentally, Greatbatch misread the color coding and installed a resistor that was the wrong size. What Greatbatch heard next astounded him—the circuit produced intermittent electrical pulses just like a human heart.

"I stared at the thing in disbelief," Greatbatch later wrote in an account of the pacemaker's development.

Greatbatch quickly realized that his tiny device could power a human heart. Heart surgeons were not so quick to embrace it. Most existing devices at the time that regulated heartbeats were the size of a television and painful for patients to use, although a smaller, external device that hooked onto a patient's belt had been developed that year. There had been some attempts to implant pacemakers in the past, but these had not proved durable.

Greatbatch's invention caught the attention of Dr. William Chardack, a surgeon at Buffalo's Veterans Administration (VA) Hospital. Chardack and Greatbatch, assisted by Greatbatch's wife Eleanor, worked on refining the device, shrinking it further in size and experimenting with various materials that would shield it from body fluids. On May 7, 1958, Chardack, assisted by Dr. Andrew Gage and with Greatbatch attending, implanted Greatbatch's pacemaker into a dog at the VA hospital. All three wore bow ties and would be later nicknamed the "bow-tie team."

"Well, I'll be damned," said Chardack as the pacemaker began to work in synchrony with the dog's heart, delivering regular shocks that forced heart muscles to contract and pump blood.

Greatbatch, a religious man, believed he was doing God's work. "I seriously doubt anything I will ever do will give me the elation that I felt that day when a two-cubic-inch electronic device of my own design controlled a living heart," he wrote.

Additional testing on animals continued until 1960. But in June of that year, Chardack implanted the device into a seventy-seven-year-old man who lived for another eighteen months before dying of natural causes, according to the National Library of Medicine. By year's end, the Buffalo team had installed ten pacemakers, two of them in children, in trials. Within a few years, artificial heartbeat procedures were widespread, saving the lives of thousands. The *Saturday Evening Post* in 1961 called it a "transistorized spark of life."

Greatbatch eventually settled on lithium batteries as a power source, a move that allowed internal pacemakers to operate for a decade. (A nuclear-powered internal pacemaker that would last twenty years was developed in 1972. Unsurprisingly, it didn't catch on due to regulatory hurdles.) Greatbatch died at the age of ninety-two in 2011, after a lifetime of tinkering that garnered hundreds of patents. The National Academy of Engineering called Greatbatch's accidental heartbeat one of the great achievements of the twentieth century. Pacemakers, meanwhile, continue to evolve in their sophistication with features like telemetry monitoring, programmability, improved connectors that reduce the chance of infection, and additional sensors.

In 1960, another great achievement in the mending of broken hearts was the first coronary bypass surgery by Dr. Robert H. Goetz, a procedure that is now a commonly performed cardiac surgery with over two hundred thousand done annually in the United States alone. Incredibly, Goetz's achievement was buried at the time. Goetz's "pioneering work was not appreciated and fell into oblivion," noted the American Association for Thoracic Surgeons in 2000, hoping to rectify a dismissive slight issued by the organization circa 1968.

Goetz's arrival in New York City, where he took up residence at the prestigious Albert Einstein College of Medicine, was circuitous to say the least. Goetz was born in Germany, the son of sculptor, and studied medicine at university in Frankfurt am Main while also working at a local hospital. While Goetz was not Jewish, he was a strident anti-Nazi. Denied a diploma and a medical license for his political stance, Goetz saw the writing on the wall.

"I felt that I better get out of Germany," Goetz later wrote in a 1999 letter to a colleague, Igor Konstantinov, a doctor at the Mayo Clinic. "I left Germany for Switzerland with ten marks in my pocket and no papers to show that I was a qualified physician."

Fortunately, a former professor, also now teaching in Switzerland to escape the Nazis, vouched for him and he obtained a medical degree in 1936. Still, Goetz was nervous about Adolf Hitler and his intentions. "Who knew when it would occur to Hitler to incorporate the German-speaking Swiss into his Great German Reich?" Goetz worried as Germany violated the Treaty of Versailles and reoccupied the Rhineland. Jews and Romani lost their right to vote in German elections that same year.

Goetz decamped to a research position at Edinburgh University in Scotland but was ultimately denied refugee status after a year by the British government. Newly married to a fellow physician and facing imminent deportation, Goetz jumped at an opportunity to work at the Groote Schuur Hospital in Cape Town, South Africa, even though it meant starting from scratch. While technically an enemy alien holding a German passport whose movements were severely restricted in South Africa during World War II, Goetz was able to hold a university teaching position but wasn't allowed physician credentials until 1944. Goetz's old German university gave him his medical diploma in 1977.

In the 1950s, Goetz drew international acclaim for his study of blood circulation in giraffes—their ability to lift their head from the ground to their full height without fainting confounded the medical profession at the time. Goetz also began researching the possibility of cardiac bypass surgery. (One of his students, Christian N. Barnard, performed the first heart transplant operation in 1967.)

Goetz, however, was increasingly at odds with the racist apartheid policies of the South African government. Once again, Goetz's political beliefs made him move, this time to New York City. By now, Goetz had worked out a possible procedure and developed the medical tools—most critically a modified Payr's cannula made of tantalum—for a cardiac bypass operation. Tests on dogs over a six-month period proved its efficacy.

On May 2, 1960, Goetz and a small team of assisting doctors performed the first coronary bypass operation on a thirty-eight-year-old New York City taxi driver. Goetz, thanks to his practice on dogs, performed the operation in seventeen seconds. The patient lived until June of the following year which, by any medical measuring stick, should have been considered a success. But Goetz's first cardiac bypass operation was his last.

Goetz was way ahead of his time. Goetz's coronary bypass procedure created a great deal of anxiety in the conservative medical profession. Resistance developed quickly as few understood what Goetz had done while other doctors vehemently opposed it. Goetz was sidelined with a transfer to vascular surgery, his accomplishment assigned to oblivion. The hospital chart, angiograms, specimens, and photographs disappeared, never to be found, noted a report by the National Library of Medicine in 2002, although a partial autopsy report of the taxi driver appears to have survived. "Resentment persisted not only in the cardiology department but in the surgery department," noted the report's analysis. Goetz himself noted that "medical colleagues were violently against the procedure" in his 1999 letter to Konstantinov, much of which appeared in a 2000 article by Konstantinov in the *Annals of Thoracic Surgery.*

Detractors became so numerous and influential that a team of doctors who performed a cardiac bypass procedure in 1964 didn't report it until 1973. Goetz's groundbreaking procedure wouldn't be referenced in medical literature for years. Credit for the first cardiac bypass operation would mistakenly be given to a Russian surgeon, Vaselii Kolesov, who performed the procedure in 1964 even though Kolesov acknowledged Goetz in his report, published in Russian, the following year. Goetz's pioneering role was lost in translation.

Goetz, however, was still all heart, subsequently inventing an intra-aortic balloon pump that gave mechanical support to patients with severe heart problems after heart attacks or operations. Goetz and his son Lionel, then a medical student at Einstein, received a patent in 1972. After his retirement in 1982, Goetz bred prize-winning bulls at a farm in Germantown, New York. Goetz died at the age of ninety in his home in Scarsdale, New York, on December 15, 2000.

Dogs also would come out ahead in the end. In the twenty-first century, canines get life-saving pacemakers and coronary bypass surgery too.

21

New York Traffic Inspires the
Maglev Train (1966)

Figure 21. A maglev train coming out of the Pudong International Airport. *Source*: Public domain.

There's got to be a better way. If you've ever driven across New York City's Throgs Neck Bridge linking the Bronx and Queens over the East River, you know that traffic can be heavy, bringing vehicles to a virtual standstill. No doubt many drivers allow their minds to daydream about somehow avoiding the stop-and-go moment they find themselves in and getting to their destination faster. But when you're a scientist at Long Island's Brookhaven National Laboratory, a research arm of the United States Department of Energy, a daydream turns into something concrete. What that dreamer couldn't anticipate is that while other nations would embrace his idea for faster travel, his own country would not.

Traffic across the bridge was no better in 1960 than it is today. Dr. James Powell was en route to Boston but had been stuck in a traffic jam at the bridge for hours. Suddenly, Powell was struck by an idea. Wouldn't it be cool to be able to somehow levitate above the traffic and arrive at a destination more quickly?

"I thought there must be a better way than driving," recalled Powell in a 1994 interview with the *New York Times*. "I get my ideas in two places: cars and bathtubs."

For most people, the idea of levitating over traffic was magical thinking. But not Powell—he figured superconducting magnets could do the job. Superconducting magnets are electromagnets cooled to extremely low temperatures to increase the strength of the magnetic field. Powell shared his idea with colleague, housemate, Canadian transplant, and semi-professional hockey player Dr. Gordon T. Danby who enthusiastically embraced the notion. Both men already were attracted to magnets. They had used magnets to design the Alternating Gradient Synchrotron, what was then the most powerful particle accelerator in the world. The partnership between the two, with Danby just two years senior to Powell, was crucial. "If we had not been working on it together, I doubt it would ever have happened," said Powell.

In their spare time, Powell and Danby developed a way to transport people and cargo using a technology with a magical name: magnetic levitation. The basic idea was that powerful magnets would suspend a train inches above elevated guideways and propel it at speeds in excess of three hundred miles an hour. The idea of a maglev train had been around for a long time—Robert H. Goddard, who would later achieve fame as the inventor of the liquid fueled rocket used for space launches, proposed the idea as a graduate student in 1909. But it was Powell and Danby's addition of superconducting magnets that made the technology powerful enough to levitate heavy passenger and freight cars. The rails for the maglev trains have two sets of connected metal coils wound into a figure-eight pattern to form electromagnets with the superconducting electromagnets, called bogies, on the train itself. As the train moves forward, initially on rubber wheels, the two sets of magnets interact. As the train's speed increases, the magnetic force becomes strong enough to lift the train and travel at higher speeds due to the lack of friction.

Maglev trains had numerous advantages that are still relevant. In addition to the very high cruising speed, maglev trains were emission-free, made very little noise compared to traditional trains, had a minimal impact on the surrounding landscape, were generally safer as derailments were unlikely, and required little maintenance since they used friction-less drive technology. Maglev trains, believed Powell and Danby, were the future of transportation

In 1967, Powell and Danby received a United States patent on their idea. Both men expected the United States government and other American investors to beat a path to their door to discuss plans for building a floating train. To their surprise, it was the Japanese who first arrived in a caravan of limousines at the rustic Brookhaven campus.

Other countries also seized upon the idea. The United Kingdom built a two-thousand-foot maglev line at Birmingham's airport, but it only traveled at twenty-six miles an hour and eventually shut down due to reliability and design problems. Other test tracks with relatively short distances sprung up in Europe. Germany experienced some successes but a three-station line in West Berlin ceased operation with reunification and interest fizzled out after a 2006 maglev train accident, caused by human error, on a test track operated by a company called Transrapid, killed twenty-six people. Russian maglev ambitions were derailed by earthquakes, war, and financial problems.

None rivaled the ambition of Asian countries where interest in maglev trains remains high. While South Korea has a low-speed airport maglev shuttle, Japan and China took a major interest in developing the fastest maglev trains on the planet.

The German company Transrapid had fortunately signed a contract prior to the deadly accident in Germany to develop a maglev train for Shanghai. In 2004, the Shanghai Maglev Train became the first commercially operated high-speed line in the world. The Chinese maglev train operates at speeds of 268 miles an hour and makes a nineteen-mile journey between the city center and the airport in seven minutes. The Shanghai Maglev Train was built for a reported $1.3 billion and when the German company declined to share the technology, cost estimates to expand maglev service rose dramatically. China's current high-speed-rail plans employ more conventional technology for running speeds of between 250 and 300 miles an hour. Still, China hasn't ditched maglev trains entirely. In 2021, China's Railway Rolling Stock Corporation tested a maglev bullet train with a speed of 373 miles an hour, potentially filling a middle ground between conventional high-speed trains and airplanes. The competition is stiff: France's TGV POS electrified train has recorded a speed of 357 miles an hour. China also operates a few low-speed, short-haul maglev trains.

Japan also is invested in "super" high-speed maglev trains, having studiously researched the technology since the first meeting with Powell and Danby. Japan ran its first test track in 1972 and currently operates a low-speed maglev train called the Linimo. In 2015, a maglev train operated by the Central Japan Railway Company broke the train-speed world record with a speed of 374 miles an hour. Starting in 2027, Japan aims to run the Chuo Shinkansen maglev train between Tokyo and Nagoya, distance of 161 miles, in forty minutes, with plans to expand to Osaka soon afterward.

In 2013, Japan offered then President Barack Obama to build the first forty miles of a maglev train in the United States for free, perhaps in acknowledgment of the key research done by Powell and Danby. By 2022, American maglev train proposals, like a Northeast Maglev route between New York City and Baltimore (using Japanese maglev technology), remained perpetually under study with the high cost of building a dedicated train line that doesn't use existing track infrastructure proving to be a major stumbling block. More puzzling were the concerns of Baltimore officials about the environmental impact of an emissions-free transportation system. Overall, American long-distance passenger trains are going in the wrong direction. According to Amtrak 2022 data, the journey of its Silver Star train from New York to Miami took 31½ hours, which is four hours longer than the same ride in 1954. Within New York state, a trip from Penn Station in Manhattan to Albany takes 2½ hours traveling at fifty-seven miles an hour, which is twice as long as a similar trip between Paris and Brussels. In Florida, a new train service called Brightline creeps toward speeds of 125 miles an hour as it emphasizes the "nicest" ride rather than fastest. And in California, a high-speed train line between San Francisco and Los Angeles has been mired in delays of various kinds and is not expected to be operational before 2030, if then.

Despite never making any real money from their idea, Powell and Danby never gave up on maglev trains. Both men worked the lecture circuit and wrote books on the subject. The pair were awarded the Benjamin Franklin Medal in Engineering in 2000 for their maglev work. The two scientists started the MAGLEV 2000 of Florida Corporation and the Danby Powell Maglev Technology Corporation to help get maglev off the ground in the United States. Still, federal funding never arrived and an effort to build a privately funded twenty-mile track in Florida stalled.

The Pied Pipers of maglev were often exasperated. "One thing that frustrates Jim and me is that we go and give talks on our whole concept and it's invariably well-received," Danby told the *New York Times* in 1994. "And yet the people who judge these things have already made up their minds. It's not that they debate it. They're like generals from the last war; they don't have to listen."

What irritated them most was the criticism that the pair were nutty professors. "We're not ivory-tower academics," said Danby. "We're used to big projects. We know how to do big projects."

They were never shy about thinking big. Powell took the maglev idea into space when, working with another former-Brookhaven colleague and aerospace engineer George Maise, authored a proposal to NASA for a "StarTram" that would use maglev technology to launch passengers and cargo into space. StarTram would accelerate a capsule, first through an eighty-mile tunnel, possibly to be built in Antarctica, and then along a tubular structure into orbit. While NASA viewed the proposal as technically sound, the projected cost was deemed prohibitive.

Danby eventually visited Japan to see a maglev train in action. "It was very nice," Danby told the *New York Times*. "It went by with a whoosh." Danby passed away on Long Island in 2016 at the age of eighty-six. Powell retired upon Danby's death, passing three years later.

While some may see Powell and Danby's pursuit of maglev technology somewhat unfulfilled, it appears maglev is a technology determined to find its way. The path forward is still murky, however. Elon Musk, best known for his Tesla electric car and SpaceX orbital endeavors, is perhaps the spiritual heir to Powell and Danby. In 2013, Musk penned a fifty-eight-page "Hyperloop Alpha" paper that resuscitated maglev interest. The key difference in Musk's concept is that the maglev train or capsule operates within a tube to eliminate air resistance. Vacuum pumps located every six miles would suck air out of the enclosed track, enabling speeds of over six hundred miles an hour, although actual speed may depend upon whether a full or partial vacuum can be sustained. As envisioned, a journey from Los Angeles to San Francisco would take forty-five minutes at a price of one hundred dollars, for example.

Two companies, Virgin Hyperloop and HyperloopTT, quickly seized the lead in hyperloop development. Virgin Hyperloop successfully demonstrated a two-seat passenger capsule on a test track in November 2020 with the promise of modules carrying twenty-eight people per trip as a goal. Nevertheless, the vision of people being whisked along at incredible speeds appears have fallen by the wayside, at least in the near term. Critics noted that hyperloop passenger capacity would still be lower than traditional trains and passenger safety is still a question mark.

Both Virgin Hyperloop and HyperloopTT now favor developing a cargo-carrying capacity first, as does another Dutch player called Hardt, which has received funding from the European Union. Hardt, founded by students from Holland's Delft University of Technology, attracted European-Union funding with an advanced switching technology that would make routing capsules easier at high speeds. The train-loving

European Union sees hyperloop technology, to be built alongside existing highways, as a way to get as many as eleven hundred trucks per day off the road as it pursues a carbon-neutral environmental agenda. Hardt estimates it will be able to move twenty-thousand half-pallets of cargo per hour at 430 miles an hour. On paper, freight makes more sense as cargo more easily tolerates faster speeds and pods can be preloaded hours before departure. More speed may be appealing to countries that have experienced supply chain disruptions. Regulatory approval also appears easier. On the downside, moving freight is not as financially lucrative as moving people.

It may be 2030 before any hyperloop plans come to fruition, say industry analysts. In April 2022, Elon Musk said his tunneling specialists at the Boring Company would develop a full-scale hyperloop test track, building off an atmospheric-pressure tunnel Musk built in Las Vegas that transports people in Tesla cars and making good on his 2013 paper at the same time. China, however, got the jump on Musk in October 2022 when scientists at China's North University tested a maglev train that operated inside a "low vacuum tube" on a 1¼-mile test track at a speed of 80 miles an hour. The goal is to construct a thirty-seven-mile test rack for its "vactrain" with a target speed of 621 miles per hour. Chinese scientists also are experimenting with installing maglev tech on passenger cars with an eye toward potentially increasing the range of electric vehicles.

Hyperloops may be most attractive in areas that do not have an existing mass transit train system in place. Powell and Danby would likely see hyperloop technology as an evolution of their own maglev concept but perhaps ruing the fact that, to this day, you still have to drive across the Throgs Neck Bridge. That's not going to change anytime soon.

22

A Queens Nurse Invents the
Security Camera (1966)

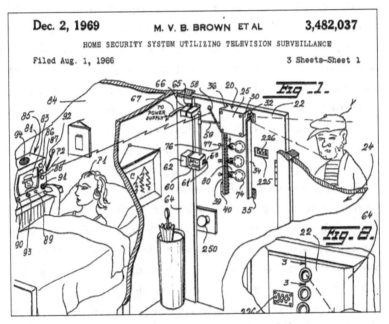

Figure 22. Patent drawing for Marie Van Brittan Brown's home security system. *Source*: United States Patent and Trademark Office, US Patent 3,482,037, filed August 1, 1966, and issued December 2, 1969.

There are at least one billion security-camera systems in use around the world. The first one was invented by a Black nurse concerned about her safety because, ironically, she felt the police response in her New York City neighborhood was slow. Her solution spawned a revolution in security and laid the groundwork for the surveillance economy that exists today.

Marie Van Brittan Brown was born in 1922 in the borough of Queens in New York City. She was the only child of parents who originally hailed from Massachusetts and Pennsylvania. Brown grew up to become a nurse, married Albert Brown and had two children.

Brown lived in the Jamaica section of Queens, and it wasn't the safest of neighborhoods in 1966. The crime rate was high and police response was often slow in an emergency. Serious crime had increased in Queens by 32 percent between 1960 and 1965. Brown's work schedule meant she would often have to come home well after dark. Likewise, her husband's job as an electronics technician meant he wasn't always at home when she arrived. In a high crime area, "Who's there?" is a question loaded with anxiety when there is a knock at the door in the middle of the night. Concerned about her own and her family's safety, the forty-three-year-old Brown decided to do something about it.

Enlisting her husband's technical expertise, Brown devised a small cabinet that affixed to the inside of the door. A motorized camera would be able to peer through a series of peepholes aligned so that people of different heights could be viewed by toggling the camera up and down. Camera images were wirelessly streamed using a radio-controlled system so pictures were available on any television in the home.

As innovative as this initial setup was, Brown knew that her home-security system needed some added features to ensure her safety. Brown added a way to open the door remotely as a further protective measure. But she was still worried about the slow response time by police, so she devised a method to contact police and other emergency services at the touch of a button.

Brown and her husband filed a patent for a "Home Security System Utilizing Television Surveillance" in 1966 and cited only three previous patents—the invention of television in 1939, and identification system developed in 1959, and a remotely controlled scanning system in 1966. No one was even thinking about a closed-circuit television (CCTV) security system at that time. Previous work in the field was limited to a German engineer named Walter Burch who had developed a remote camera–monitoring system to monitor Nazi V-2 rocket launches in World War II. Brown and her husband were granted a patent in 1969. The issuance of the patent caught the attention of the *New York Times* who interviewed her that same year. "With the patented system, a woman alone in the house could alarm the neighborhood immediately by pressing a button and installed in a doctor's office it might prevent holdups by drug addicts," the newspaper noted. The accompanying photo of Brown and her husband shows a woman of remarkable intensity.

When asked about a next step, Brown told the *Times* she intended to explore manufacturing options and supply home builders. That commercial development never happened as at that time the cost of Brown's home security system proved prohibitive. Brown faded from view and died in Queens in 1999 at the age of seventy-six.

Brown clearly was ahead of her time. As the components became less expensive over the ensuing years, Brown's vision became realized. By 2005, home-security systems were becoming widely available to residential customers. CCTV surveillance is a multibillion-dollar business. Brown's pioneering work made it all possible and has been cited in dozens of subsequent patents. It is the foundational inspiration for products like the Ring doorbell that relays images from the front door to your phone. Brown's invention also has been recognized with an award from the National Scientists Committee. Perhaps most rewarding of all for Brown is that she inspired her daughter Norma to also become a nurse and an inventor with several patents issued to her.

Brown's real legacy is much more impactful. CCTV security systems are prevalent across all walks of life. Cameras controlled by computers can track, identify, and categorize objects seen through their lenses. This ability is now so sophisticated that it can identify a person moving through a crowd in a direction not normally expected, like a person moving away from a crowd entering a sports stadium. And these individual cameras can be networked together for coverage of vast areas—entire cities in fact. This development is a source of controversy as critics cite civil liberties concerns and the loss of privacy in public spaces.

Brown's initial desire to see who was at her front door without opening it has morphed beyond expectation. Internet protocol (IP) cameras allow camera owners to view images remotely so you can see who is at the front door even when you're not home. Cameras likewise have become smaller, wireless, and more sophisticated with high resolution; motion detection; zoom, pan, and tilt; and multiple-sensors capabilities. Some CCTV systems even have the ability to "talk" to offenders.

More controversially, home-security systems now offer face recognition technology that allows homeowners to build their own database of people authorized to be on the premises. Networked linkage to police takes Brown's push-button alarm to a new level. On a larger scale, facial recognition is being used as an identifier in public spaces like airports and may even be used to unlock your mobile phone. Critics worry that facial recognition is potentially an abusive tool in the hands of authoritarian-minded governments looking to monitor their citizens. While some municipalities have banned facial recognition technology citing privacy concerns, it's conversely becoming more widely used in specific settings and applications. Still, facial recognition isn't foolproof—one man stole money from his girlfriend's bank account by pulling up her eyelids while she was sleeping to activate the phone's facial recognition software. Facial recognition has been criticized for having higher identification error rates for darker-skinned individuals. Facial recognition technology also has spawned a small industry of anti-CCTV countermeasures like "reflectacles" sunglasses that reflect infrared and visible light to make the wearer appear as a white blur to cameras. Brown's basic answer to the question of who was at her front door is one that everyone around the world now has the ability to know in detail.

23

Break This

The Birth of Hip-Hop (1973)

Figure 23. Afrika Bambaataa. *Source*: Arturo Almanza, Creative Commons Attribution–Share Alike 4.0 International license.

New York has always been a hotbed for music. If there are eight million stories in the Naked City as the 1950s New York police procedural television series claimed, then there seems to be a song for every one of them. The number of musicians who have sung about New York City would fill every seat in a large concert hall. Frank Sinatra's "New York, New York" is arguably the best-known song about the Big Apple.

Other vote-getters for best song about New York City by famous artists, many with New York roots and all from notably widely divergent cultural backgrounds, would be tunes by Billie Holiday, the Pogues, Billy Joel, Sting, Jimmy Cagney, Leonard Cohen, Joni Mitchell, the Ramones, Jay-Z and Alicia Keys, Bob Dylan, Irving Berlin, Lady Gaga, Ed Sheeran, Jennifer Lopez, Nat King Cole, Duke Ellington, Simon & Garfunkel, and Lou Reed. "American Pie," considered a masterpiece and voted as one of the top five songs of the century by the Recording Industry Association of America and the National Endowment of the Arts, is largely about the youthful musings of Don McLean as a teenager in New Rochelle in Westchester County.

Conversely, Taylor Swift's "Welcome To New York" was pegged the worst song ever about New York by the *Gothamist,* a local news and cultural website. More deliberately, some musicians have penned songs that are less celebratory like Ray LaMontagne's "New York City's Killing Me" or explore a more complicated love/hate relationship like "New York, I Love You but You're Bring Me Down" by LCD Soundsystem.

Meanwhile, New York music venues ranging from the Cotton Club to CBGB, gained national prominence, following in the footsteps of the Armory Hall, a dance hall owned by former Five Points gang member Billy McGlory. Armory Hall put on "one of the wildest and most shameless revels which has ever disgraced the record of local occurrences," sniffed the *New York Times* in 1883, reporting on a music genre that would become known as ragtime.

Beyond individual songs, New York has nurtured a wide range of music genres ranging from jazz to musical theater to punk. Some genres, like salsa, were born in New York City and continue to have its adherents. Others, like disco, shone brightly and then flamed out. But perhaps no native New York musical genre has risen from obscurity to worldwide influence with the staying power of hip-hop. And a good deal of its success can be linked to music streaming technology that revolutionized music distribution.

The origin of hip-hop, also known as rap, is well documented. On August 11, 1973, eighteen-year-old DJ Kool Herc—real name Clive Campbell—began hosting neighborhood parties from his apartment building at 1520 Sedgwick Avenue in the Bronx. DJ Kool Herc's key innovation was to use a two-turntable setup to isolate the drumbeat, the "break," of the same song on both turntables and elongate it by switching back and forth between them. The technique became known as "breaking" or "scratching."

DJ Kool Herc was an immigrant familiar with Jamaican parties that used massive sound systems. The Jamaican disc jockey custom of "toasting" led to DJ Cool Herc's exhortations to dancers that became known as rap. DJ Cool Herc's first sound system comprised two turntables, two amplifiers, two speaker columns, and a PA system. DJ Kool Herc developed the early techniques that defined hip-hop like the "merry-go-round" switching from one break to another and the rhyming style of talking in slang phrases to the audience like "This is the joint! Herc beat on the point!" According to music journalist Steven Ivory, hip-hop began when DJ Cool Herc placed two copies of James Brown's 1969 *Sex Machine* on each turntable and ran an extended mix of the percussion segments on the song "Give It Up or Turnit a Loose."

DJ Cool Herc quickly became a folk hero in the Bronx and began playing nearby clubs, high schools, street parties, and parks. Hip-hop events expanded to include dedicated rappers. Acrobatic break dancers that became known as "crews"—often linked to individual artists—were soon part of the scene as was stylized graffiti that became enigmatic tags on subway cars all over the city. Crews would have tight links to individual rappers.

Hip-hop was a welcome development in poor neighborhoods. In the 1970s, the Bronx was burning due to a combination of arson and misguided city planning that closed down key fire houses, allowing blazes to career out of control. Over the course of a decade, 40 percent of the buildings in the South Bronx were burned or abandoned, according to some estimates. Crime flourished and street gangs turned some areas into no-go zones for the authorities. The New York City economy was stagnant and unemployment rates were high. Hip-hop turned more than one person away from crime to a life in the arts. Afrika Bambaataa, a former general in the Black Spades street gang who transitioned to a peace-loving Zulu Nation and became known as the godfather of hip-hop is perhaps the most famous example. At times, the line blurred. In the Blackout of 1977, DJ sound systems were reportedly the most looted item.

"If blues culture had developed under the conditions of oppressive, forced labor, hip-hop culture would arise from the culture of no work," observed author Jeff Chang in *Can't Stop Won't Stop: A History of the Hip-Hop Generation.*

Hip-hop was a street-level view of urban poverty. "Street smarts" were a measure of credibility and authenticity. One of the most famous lyrics came from DJ Cool Herc's contemporary Grandmaster Flash and the Furious Five: "I can't take the smell, can't take the noise/Got no money to move out, I guess I got no choice/Rats in the front room, roaches in the back." Grandmaster Flash rapper Keith Cowboy is often credited with coming up with the hip-hop name that defined the genre. Raps were often very political in their messaging. Critics chided hip-hop for its aggressive lyrics, chauvinism, and as a potential flashpoint for civil unrest.

"The music is meant to be provocative—which doesn't mean it's necessarily obnoxious," explained New Yorker Jay-Z, one of the stars of what's collectively called the Hip-Hop Nation. "But it is (mostly) confrontational and more than that, it's dense with multiple meanings. Great raps should have all kinds of unresolved layers that you don't necessarily figure out the first time you listen to it."

Meanwhile, most of America ignored what was happening in the Bronx, per usual. But there was headway. Hip-hop began to become more than just a music genre, transforming into a culture with its own vocabulary and expanding first to Philadelphia and then across the country. In 1979, "Rapper's Delight" by the Sugarhill Gang, released on an independent African-American owned label, got radio play and became a chart-topping phenomenon that caught America's attention. Artists like Debbie Harry of Blondie fame began to incorporate hip-hop elements into their music. Still, there was resistance. MTV debuted in 1981 and turned music videos into an entertainment but the channel steadfastly ignored hip-hop for years. It wasn't until 1986 when Queens-based Run-DMC exploded onto the channel with three hit videos over two years, one with rock band Aerosmith, that hip-hop became acceptable.

Run-DMC ushered in what many consider to be a golden age of hip-hop as the genre reached out to a wider audience. Among the noteworthy groups of the era were the Beastie Boys, a white trio who popularized digital sampling. Female singers like New Yorker Nicki Minaj signaled an alternative to rap's often misogynistic viewpoint. *Straight Outta Compton* by Los-Angeles-based N.W.A challenged New York's ascendant role even as the Big Apple's Notorious B.I.G. cemented a reputation as the greatest rapper ever. The trash-talking East Coast versus West Coast rivalry would turn deadly with the murders of hip hop stars Tupac Shakur in 1996 and Notorious B.I.G. the following year.

New York, of course, was a central theme of many rappers with an address there. Visions of the city range from darker looks like "N.Y. State of Mind" by Nas to more exuberant tracks like "No Sleep Till Brooklyn" by the Beastie Boys and even exhilarating holiday raps like "Christmas in Hollis" by Run-DMC but hip-hop's reach was already expanding beyond New York and its Bronx origins.

Hip-hop's reach extended across the globe to embrace many cultures and themes. Rappers from the United Kingdom to Japan made their voices heard. MTV began airing its "Yo! MTV Raps" show that gave hip-hop artists wider exposure. Sub-genres like gangster rap, party rap, and political rap appeared. In the United States, the South joined the East and West Coasts in developing a distinctive regional sound. Meanwhile, on the recording side, innovations like the pitch-altering auto-tune device developed by Antares Audio Technologies were widely adopted to correct out-of-key voices.

Social media and blogs helped hip-hop garner more fans. Hip-hop jumped on the back of digital music streaming technology and never looked back. Streaming services like SoundCloud, Apple Music, and Spotify changed the music distribution model. Hip-hop capitalized on it as the genre's short song format translated into more revenue under a pro rata payment scheme that pooled revenue and paid out on the share of total tracks, a model that rewarded short tunes, according to a 2022 study by researchers at the University of Hamburg and Kühne Logistics University in Germany.

Hip-hop artist Kanye West, admittedly a big ego in a genre full of them, declared his 2013 album *Yeezus* marked the death of the compact disc format. In 2017, the Canadian hip-hop artist Drake released a streaming-only project, which he called a "playlist," insisting that it was neither a mixtape nor an album. Drake's 2018 *Scorpion* release garnered five hundred thousand CD sales versus six billion on-demand streaming downloads, according to Buzzfeed, the New York news website. By then, hip-hop had usurped rock and roll as the most consumed genre of music.

24

The Body Scanner (1977)

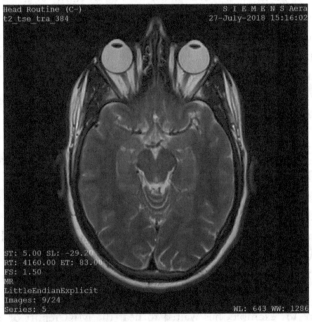

Figure 24. MRI image of a brain. *Source*: Ptrump16, Creative Commons Attribution–Share Alike 4.0 International license.

The New Yorker was robbed. That was the consensus view when Dr. Raymond V. Damadian was denied the Nobel Prize in Medicine for his work in the development of magnetic resonance imaging, commonly known as MRI. Most authorities acknowledge Damadian as the inventor of MRI so when the Nobel Prize recognizing "discoveries concerning magnetic imaging," was awarded, Damadian's absence was widely noticed. Two scientists were named as prize winners. Nobel rules permit the naming of as many as three individuals as recipients. The careful wording of the award was interpreted as meaning the omission was deliberate.

Why the snub? There's lots of speculation. Damadian was a physician rather than a scientist, so he wasn't in the "club" so to speak. Others point to less-than-endearing personality that garnered him more enemies than friends. And perhaps most damning of all was that Damadian was a Bible-loving creationist who believed the Earth was created six thousand years ago in seven days. If Damadian were awarded the 2003 Nobel Prize in Medicine it would be viewed as an evolution-denying endorsement of creationism, noted observers. Nobel deliberations are kept secret for fifty years so it will be a long time before we're privy to their thinking.

Damadian, a New Yorker through and through, wasn't the sort to take the snub lying down. Full-page ads quickly appeared in the *New York Times, Washington Post,* and *Los Angeles Times.* Dr. Damadian declared that he had taken the first steps in developing magnetic resonance for medical scans in the 1970s and that he should be recognized for his work. The ads, which cost hundreds of thousands of dollars, pictured a Nobel Prize upside down. Damadian declared the omission "the shameful wrong that must be righted."

Damadian ultimately got plenty of recognition elsewhere and the enduring gratitude of patients with life-saving diagnoses after MRI scans. Originally, developed as a way to spot cancer, MRI technology is now more widely used for a variety of medical purposes. Nearly every sport fan, for example, has waited to hear the MRI report of a favorite athlete suddenly injured. The number of MRIs performed annually in the United States alone is estimated to be approximately thirty million. Even dogs and cats now get MRIs.

Damadian being dissed by the Nobel Prize Committee might have been half-expected as his entire career is one that can be described as an underdog battling against the odds. It's not for nothing that Damadian named his first MRI machine "Indomitable."

Damadian was born in Manhattan in 1936 but was basically a Queens guy, growing up in Forest Hills. Damadian's father was a newspaper photo engraver who had fled the Armenian genocide in 1915 at the age of twelve, arriving in New York City after World War II where he married his French-Armenian bride Odette, an accountant. Young Damadian was soon recognized as a child prodigy, attending the famous Juilliard School of Music on weekends from the age of eight where he played violin. That was in addition to his normal academic career at public schools in Forest Hills. With the historic West Side Tennis Club in Forest Hills in easy proximity, Damadian developed the skills of a good tennis player. Damadian would lose his beloved grandmother to cancer—her death pointed him toward a career in medicine.

At fifteen, Damadian won a Ford Foundation scholarship that allowed him to enter university without having to graduate high school first. The award had a draft-style admissions procedure and Damadian found himself enrolled at the University of Wisconsin. Damadian's arrival did not go unnoticed: "Percival Suckthumb Arrives" the school newspaper headlined with an illustration of a Lord Fauntleroy figure in prepubescent curls. While the pointed barbs were blunted by a thick skin developed in New York, Damadian later admitted the negative attention left him "feeling different." Damadian got his degree in mathematics in four years and scampered home to New York City to enroll in the Albert Einstein College of Medicine. By the 1970s, Damadian had military service in the United States Air Force during the Vietnam War behind him and was a successful physician at the SUNY Downstate Medical Center in Brooklyn. A persistent abdominal pain and the memory of his grandmother got Damadian thinking of a way to view the body's interior for early detection of disease.

The body pain disappeared of its own accord, but Damadian's research continued. In 1937, Columbia University professor Isidor. I. Rabi observed the quantum phenomenon dubbed nuclear magnetic resonance (NMR). Rabi recognized that atomic nuclei reveal themselves by absorbing or emitting radio waves when exposed to a strong magnetic field. Damadian, using a primitive nuclear magnetic resonance machine, a tool then commonly used by chemists, turned it on rats implanted with tumors. Damadian discovered that the difference between healthy and cancerous tissue was noticeable. The key marker was hydrogen. The tumors held more water, so the signal was different than that of healthy tissue. Once the barrage of radio waves was turned off, Damadian found that telltale emissions from the cancerous region lingered or "relaxed" longer. Damadian published an article on his findings in *Science* in 1971, suggesting the technology could be used to treat cancer, heart and kidney disease, as well as mental illness.

It wasn't well received. NMR machines were small and spun test samples at speeds of up to one hundred rotations per minute. Making one large enough to fit a human seemed farfetched. "How fast are you going to spin the patient?" asked one wag.

Damadian was undeterred. "When the idea is compelling enough, it can be difficult to evade its allure," Damadian later wrote. "It seems to repeatedly force itself into consciousness to remind you of your failure to act."

Damadian managed to get a small grant from the National Cancer Institute to continue his research, but more money was needed. With little interest from the NMR community, which seemed more enamored of concurrent NMR imaging research conducted by physicists, Damadian relied on private funding solicited by his brother-in-law. In 1974, Damadian was granted a patent for his "Apparatus and Method for Detecting Cancer in Tissue." Wisely, he dropped the term "nuclear" from the machine's name, reasoning it would most likely frighten future patients. Now he just had to build one.

Damadian, with a postdoctoral support team comprised of Larry Minkoff and Mike Goldsmith, soon had the basic framework of a human-sized machine worked out. Virtually every segment of the machine was built from scratch. The heart of the machine would be a large super-conducting magnet they would have to build on their own with about thirty miles of wire. The team caught a lucky break when they were able to buy miles of superconducting wire cheaply from Westinghouse, which had decided to exit that business. But Goldsmith still had to shape the wire into hoop-shaped coils that would encircle the patient. Minkoff learned to weld by reading a three-part tutorial in a magazine.

It took the trio a year to build their machine with its characteristic doughnut-shaped opening for the patient. Complicating construction was that the magnets had to be cooled with liquid helium to absolute zero to function in the desired manner. And while an MRI scan is called an image, it's not actually a photograph but a culmination of magnetic fields and radio wave measurements that have to be processed. By the time they were built, the machine weighed one thousand pounds and required a tackle and hoist to move it into position. The machine was named "Indomitable."

Damadian volunteered to be the first person to be scanned, wearing a corset-like "antenna" fashioned from cardboard, capacitors, and copper wire. The moveable platform Damadian lay on would slide him into the machine. "I could just barely get into it," Damadian recounted in *Inc* magazine in 2011. "I had a cardiologist and emergency shock paddles on hand in case something went wrong." The machine didn't work. "It was a profound disappointment. We hypothesized the scan failed because, frankly, I was too fat for the coils."

All eyes landed on Minkoff. "It took some convincing but Larry Minkoff, who was very skinny, finally agreed to get in." Minkoff held out for several weeks, monitoring his boss for any adverse side effects. Minkoff's lean physique was to the machine's liking. "To our great excitement, we got a signal right away," said Damadian. "We had achieved the world's first human scan. We were ecstatic. That was July 3, 1977." The MRI of Minkoff's chest showed his heart, lungs, vertebrae, and muscular tissue. Wine bottle corks were popped in celebration.

However, major manufacturers showed such little interest in producing MRI machines that Damadian was forced to create his own manufacturing company called FONAR headquartered in Melville, Long Island. Once again, the brother-in-law was called upon to raise funds from amongst his well-heeled contacts. Damadian continued to refine the process, quickly bringing scanning time down from hours to thirty-eight minutes by the following year. Superconducting magnets were replaced by easier to maintain permanent magnets. An Italian sculptor was hired to design a more attractive shell casing. Image resolution improved. In 1980, the first commercial MRI scanner went on sale. Initial customers hailed from the United States, Mexico, Japan, and Italy. Damadian designed a mobile unit that could fit inside a large truck.

FONAR Corporation became a success as MRIs were seen as an attractive alternative to X-rays and CT scans that used radiation to yield results while also spotting tumors missed by those techniques because they were so small. But big competitors soon appeared. And while the industry collectively modified its approach to MRIs over time, Damadian was certain the competition was violating his initial patent. The journey through the courts took years but Damadian was right. In 1997, GE was hit with a $128 million judgement in favor of FONAR. Hitachi and Siemens settled out of court. FONAR continued to innovate, coming out with an upright MRI scanner for the neck and spine in 2001. Damadian also collaborated with Wilson Greatbatch to develop an MRI-compatible pacemaker. MRIs became widely used even far afield with one MRI-based study revealing that composers shut down part of their brain to create music. Smaller, more portable units also are becoming widely available.

Damadian was perhaps understandably a bit peeved when he was ignored for the Nobel Prize and not honored alongside British physicist Peter Mansfield and fellow New Yorker and chemistry professor Paul Lauterbur of SUNY at Stony Brook. (Damadian and Lauterbur were apparently not on good terms.) Even a staid publication like the *Journal of Urology* opined that "excluding Dr. Damadian seems to be a serious and purposeful omission." Damadian to his credit put the Nobel Prize issue behind him pretty quickly. In 2004, the prestigious Franklin Institute in Philadelphia gave Damadian its Bower Award for his MRI invention.

"There is no controversy in this," Dr. Branford A. Jameson, a professor of biochemistry at Drexel University who chaired the selection committee told the *New York Times*. "If you look at the patents in this field, they're his."

At the ceremony, Damadian told the *Times:* "I put that issue behind me, and I don't want to talk about it. I made the original contribution and made the first patent. If people want to reconsider history apart from the facts, there's not much I can do about that." Damadian's ego may have been partly soothed by the knowledge of other cringe-worthy decisions made by the Nobel Prize Committee such as the 1949 medal given to Portuguese Egas Moniz who pioneered the use of frontal lobotomies for psychiatric disorders.

Indomitable is housed in the Smithsonian Museum. Damadian belongs to the institution's Hall of Fame. Damadian also received the National Medal of Technology and was inducted into the National Inventors Hall of Fame, among many other accolades. Damadian died on August 3, 2022, at the age of eighty-six at his home in Woodbury, New York.

25

Solar Lights Up the Dangerous Dark (2015)

Figure 25. SolarPuff lantern. *Source*: Courtesy Alice Min Soo Chun/SolarPuff.

The amount of sunlight that strikes the earth's surface in an hour and a half is enough to handle the entire world's energy consumption for a full year, according to the United States Department of Energy. And every year, humans get better and better at harvesting sunlight for use as power. The invention of the modern solar panel dates to 1954 but since 2000, the solar-power industry has experienced soaring growth rates as solar panels become more efficient. Solar power has moved from being the darling of off-the-grid survivalists to a mainstream source of energy.

Alice Min Soo Chun is a New York inventor contributing to the increased use of solar power, albeit in a deliberately small way. Chun comes to the solar-power arena with an impressive academic background in architecture and materials technology, having taught at the Parsons School of Design and Columbia University in New York City. Chun also is the former director of the Materials Resource Lab at the New School. But the inspiration for a small sun-powered device called the SolarPuff comes from her childhood years in Seoul, South Korea.

The event that galvanized Chun was the 2010 earthquake in Haiti that displaced a million people. Chun noticed that Haitians were relying on dangerous kerosene lamps for light. The risks associated with kerosene lamps are well known. In addition to being a fire hazard, the World Bank estimates that breathing in kerosene fumes is a cause of lung cancer. Kerosene also is expensive—for many Haitians kerosene purchases accounted for 30 percent of their income. Chun knew her solar-powered device had to be inexpensive.

Chun developed some early prototypes with her students at Columbia University, but more work was needed to develop a final product that was functional yet aesthetically pleasing. Chun's thoughts harkened back to her childhood. While born in Seoul in 1965, her family subsequently moved to Syracuse in upstate New York in 1968. Chun's father was a mathematician and while her mother was an artist and textile maker. But life in Syracuse wasn't easy. "We were marginalized as the only Asian family living in an all-white neighborhood outside of Syracuse, Chun told *Authority Magazine* in 2021. "There were many days when I walked home from school with a black eye or bruises from being bullied."

But one memory proved inspirational. Chun's mother had taught her origami, the ancient art of turning a flat, square piece of paper into a finished sculpture through folding and sculpting. Origami forms, Chun realized, are a perfect marriage between mathematics and art. Chun's mother also had taught her how to sew her own clothes.

Chun began sewing solar panels onto fabric with an eye toward using softer, more malleable material as a host. A further motivating factor was her son Quinn's diagnosis with asthma, a condition often linked to air pollution. Chun views solar power as a clean energy source worth pursuing.

Eventually, the design of SolarPuff took shape around some key features that would make it easy to use. Chun invented a solar light that was portable and self-inflatable, eliminating the need for a mouth nozzle that might spread germs among multiple users (like in a family). Basically, the device has four components. One is a solar panel that captures the sunlight. A rechargeable lithium-ion battery similar to one found in cell phones stores energy. LED bulbs produce the light. Lastly, there is a recyclable polyethylene terephthalate (PET) material that inflates into cube and pyramid shapes when it is pulled open by the user. After a full eight hours of charging in the sunlight, SolarPuff provides eight hours of light in the high setting and twelve hours in the low setting. The number of hours charging largely correlates to the length of time the lantern will work, although charge times may vary depending on the season and the sun's proximity to Earth. The main limitation is temperature. SolarPuff won't recharge when it's below freezing. The battery itself will hold a charge for about a year.

But Chun needed to make sure the device worked. Chun field-tested the device over a three-year period in Haiti. Handmade SolarPuffs were distributed among rural women farmers, most of whom lived with multiple children in one-room dwellings with the kitchen outside. Vulnerable communities, especially women and children, were most at risk in the aftermath of a natural disaster when darkness falls, and the risk of kidnapping and assault was greatest.

"When I first gave them the SolarPuffs, they started to sing and dance, saying this was a gift from God," Chun told *Fast Company* in 2017. "That's when I knew I had to make this accessible to everyone."

In 2015, Chun launched a company called Solight Design to produce the lantern. A large part of the funding came from a successful Kickstarter campaign. Chun was quickly recognized for her achievement.

Among the awards Chun received was the United States Patent Award for Humanity. SolarPuff was exhibited at the Museum of Modern Art (MOMA) in New York City. "One overlooked ingredient for success is empathy," says Chun.

Since then, Solight Design has created a variety of SolarPuffs in different sizes and color-lighting schemes that can be used for everything from camping to pool lights. But the product's biggest impact clearly is in disaster relief. Perhaps Chun's biggest success was in Puerto Rico when one hundred thousand SolarPuffs were distributed through NGO relief organizations to people without electricity after Hurricane Maria in 2017. The mayor of San Juan called SolarPuff "a cube of hope."

SolarPuff is credited with helping to reduce crime in Syrian refugee camps. SolarPuffs have been distributed to natural disaster victims in Nepal, Dominica, and Columbia, among others as links to NGOs increase. SolarPuff's small size makes shipping costs manageable. Ten percent of the profits go to a charitable organization linked to the company. People also can send a SolarPuff directly to a designated partner charity via the company's website, solight-design.com.

Solar Puff often is linked to humanitarian missions in disaster zones but Chun notes that it has an everyday use that helps the environment, an idea she calls individualized infrastructure. "We don't have to tap into the grid all the time,' says Chun. "One SolarPuff used for a few hours per day instead of a regular light bulb, one person can save over ninety pounds of carbon emissions. Climate change is real, and we all have the power to make a difference to change the world."

Changing the world is what good inventors do.

Index

20th Century Limited, 88

Abolitionism, 7–10, 18
Air conditioning, 64–70
Adirondack chair, 1
Albert Einstein College of Medicine, 130, 153
Alternating Gradient Synchrotron, 135
Air Force One, 85, 89
American Colonization Society, 7
American Red Cross, 112–114
Ansolabehere, Professor Stephen, 54
Antares Audio Technologies, 149
Apple iPhone, 47
Armory Hall, 146
Army Tactical Brassiere, 82
Arthur, Chester, 9
Atomic bomb, 116
Auto-tune, 149

Ballot marking devices, 53
Bambattaa, Afrika, 145, 148

Barnum, P. T., 23, 37
Baseball, 11–16
Bath, Patricia, 2
Baldwin, William, 26
Batman, 102–108
Beastie Boys, 23
Berryman, Clifford, K., 74
Biden, Joe, 49
Black Convention Movement, 7
Blodgett, Katherine Burr, 97–101
Blood for Britain, 112
Bloodmobile, 109, 112
Blood plasma, 111
Blue Frontier, 70
Brassiere, 77–83
Brookhaven National Laboratory, 2, 134
Brown, James, 147
Brown, John, 18
Brown, Marie Van Brittan, 141–144
Brownie camera, 44
Brunot, James, 91, 93–95
Buffalo Forge Company, 65–67

Build-A-Bear Workshop, 75
Butts, Alfred Mosher, 91–96

Cadolle, Herminie, 78
Carrier, Willis, 2, 64–70
Cartwright, Alexander Joy, 11–16
CBS, 124
Chadwick, Henry, 14–16
Chang, Jeff, 148
Chardack, William, 129
Chun, Alice Min Soo, 159–161
Chuo Shinkansen, 137
Cleveland, Grover, 58
Cobb, W. Montague, 113
ColdSnap, 70
Coldspot refrigerator, 87
Collier, Holt, 74
Columbia University, 17, 18, 24,
   25
Columbian Exposition, 57
Comics Code Authority, 106
Coney Island, 35–39
Connelly, Anna, 2
Cooper, Peter, 23, 24
Corset, 78–79
Criss-Cross, 93
Coney Island, 35, 36
Corning, 2, 43
Coronary bypass surgery, 130–133
Creole, 7
Crosby, Caresse, 80–83
Crosby, Harry, 80–81
Crum's restaurant, 20
Crystal Palace, 23
Current War, 56
Cyclone roller coaster, 39–40

Damadian, Raymond V., 151–157
Danby, Gordon T., 135–140
Davis, Sylvanus, 51
DC Comics, 102, 107
Death ray, 61–63
Delamater Iron Works, 30
Dewey, Admiral, 32
Dickman, Josephine, 45
Doubleday, Abner, 14, 15
DJ Kool Herc, 147
Dominion Voting Systems
   Corporation, 53
Douglass, Frederick, 6, 8, 9
DRE voting machines, 53
Dress Reform Association, 79
Drew, Charles, 109–114
Dry cleaning invention, 5–10
Dreyfuss, Henry, 88

Eastman, George, 2, 41–47
Edison, Thomas, 44, 55–57, 59
Electric Boat Company, 32
Electrical Exposition of 1898, 59
Elevator, 22–27
Elysian Fields, 13
Empire State Building, 26
Equitable Life Building, 24
E. V. Haughwout Building, 24

Facial recognition, 22
Fenian Brotherhood, 30
Fenian Ram, 31
Fermi, Enrico, 118
Finger, Bill, 104–108
Finlay, Robert, 7
FONAR, 155

Freedman Hospital, 111, 114
Fulton, Robert, 1, 79

General Electric, 2, 15, 24, 156
George Eastman Museum, 45
GG1 locomotive, 87
Gillespie, Albert, 51
Goddard, Robert H., 135
Goldmark, Peter, 123–127
Goldsmith, Mike, 154–155
Goetz, Robert, 130–133
Gone with the Wind, 97, 124
Gotham City, 105
Grandmaster Flash, 148
Gravity Pleasure Switchback Railway, 37
Greatbatch, Wilson, 128–130, 156
Groot Schuur Hospital. South Africa, 131
Grumman, Leroy, 1
Gund, 75

Haiti, 159–160
Hall, Blakely, 31
Hardt maglev train, 139
Help America Vote Act, 52
Hemingway, Ernest, 81
Hip-hop, 145–150
Holland, John, 29–34
Holland VI, 32
Home security system, 22
Hot dog, 37
Howard University, 111
Hunley, 29
Hupmobile, 86
Hyatt, John Wesley, 1

Hyperloop, 139–140

IBM, 2, 52
Ideal Novelty & Toy Co., 75
"Indomitable," 153
Invisible glass, 98, 100
Iron Beam Defense System, 63

Jacobs, Mary Phelps, 77, 79–80
Jameson, Dr. Branford A., 156
Jay-Z, 148
Jennings. Elizabeth, 9–10
Jennings, Thomas L., 5–10
Jogbra, 82

Kane, Bob, 103–107
Kaons, 118
Kelly, FBI Director Clarence, 62
Konstantinov, Igor, 131, 132
Knickerbocker Base Ball Club, 12
Kodak, 41–47
Kolesov, Vaselii, 132
Kosanovic, Sava, 62

Langmuir, Irving, 98–101
Lawrence Livermore National
    Laboratory, 116
Lay, Herman, 20
Lee, Canada, 82
Lee, Teng Dao, 119
Lexico, 93
Liberty calendar, 46
Lindbergh, Charles, 40
Loewy, Raymond, 2, 84–90
Longino, Governor Andrew H.,
    72–73

Low vacuum tube maglev train, 140
LP record, 122–127
Luce, Clare Booth, 120
Luna Park, 35, 39

Macy's department store, 94
Maglev train, 134–140
Magnetic levitation, 134–140
Magnetic Resonance Imaging (MRI),
    151–157
Maise, George, 138
Manhattan Project, 117–118
Marconi, Guglielmo, 60
Marvel Comics, 108
Mauch Chunk Gravity Railway, 36
MAYA (Most Advanced Yet
    Acceptable), 87, 90
Medicis, Catherine de, 78
Michtom, Morris, 74–75
Michtom, Rose, 74–75
Miles, Alexander, 25
Miller, John, 39
Mills Commission, 15
Minkoff, Larry, 154–155
Myers, Jacob H., 8
Miles, Alexander, 4
Miller, John, 6
Moon's Lake House, 18
MTV, 149
Murphy. Joe "Spud," 20
Musk, Elon, 56, 139–140
Myers, Jacob H., 50

National Baseball Hall of Fame, 12,
    15
NASA, 13

New York Central Railroad, 88
New York Kidnapping Club, 7
New York Nine, 13
"New York, New York," 146
New York, songs about, 146
Nin, Anais, 81–82
Nobel Prize, 61, 63, 115, 119–120,
    151–152, 156–157
Nobleman, Marc Tyler, 103
North University (China), 140
Notorious B.I.G., 149

Oppenheimer, J. Robert, 118
OSET Institute, 54
Otis elevator, 22–27
Otis, Elisha Graves, 22–27

Pacemaker, 128–130
Parity Theory, 118
Pennsylvania Railroad, 87
Perez, Eddie, 54
Players Club, 58–59
Plunger, 32
Poe, Edgar Allan, 50, 92, 104
Powell, James, 135–140
Potato chip, 17–21
Popular Science magazine, 51
Presbyterian Hospital, 111

Rabi, Isadore I., 153
Railway Rolling Stock Company
    (China), 137
Rational Dress Society, 79
Reichenbach, William, 44
Remote control, 55–63
Rivoli theater, 68

Robin, 105
Robinson, Jerry, 105
Roller coaster, 35–40
Roosevelt, Theodore, 32, 72–76
Run-D.M.C., 23

Sackett-Wilhelms Lithographing and
    Publishing Company, 65
Saratoga Springs, 3
Sarnoff, David, 124
Sasson, Steve, 47
Scrabble, 91–96
Scudder, John, 111
Shoup, Samuel and Ransom 52
Sex Machine, 147
Sinatra, Frank, 1, 146
Singer, Isaac, 2
Smith, Gerrit, 17
Solarpuff, 159–161
Solar power, 158–16
Solight Design, 160–161
Soviet Union, 61–63
Space elevator, 27
Spalding, Albert, 15
Speck, George "Crum," 17–21
Star Tram, 138
Star Wars, 63
Strategic Defense Initiative, 63
Straus, Jack, 94
Steiff, Margaret, 75
Steinmetz, Charles Proteus, 2
Streaming services, 150
Streamline Moderne, 87
Strong, Henry, 43
Studebaker, 88–89
Submarine, 28–34

Sugarhill Gang, 149
Superman, 103, 105
Surveillance economy, 22

Tappendum, William, 21
Teddy bear, 4, 71–76
Television, 124–125
Tesla, Nikola, 2, 55–63
Thompson, LeMarcus Adna, 36–40
Turtle, 29
Trump, Donald, 49, 53, 63
Trump, John, 63
Twain, Mark, 59

United States Navy, 29, 32
Uranium, 117
U.S. Standard Voting Machine
    Company, 51
U.S.S. Holland. 33
U.S.S. Nautilus, 33
U.S. War Department, 113

Vanderbilt, Cornelius, 18, 19, 20
Vardaman, James K., 72–73
Vermont Teddy Bear Company, 75
Verne, Jules 29, 37
Vinyl records, 125–126
Voting machine, 48–54
Votomatic, 52
VTOL aircraft, 61

Walker, Madam C.J., 2
Warner Brothers Corset Company,
    80
Washington, Booker T., 11
Wells, Jonathan D., 7

Wertham, Fredric, 106
Westinghouse, George, 57–58
Wheeler, Seth, 13
Wicks, Kate, 19
Wu, Chien-Shiung, 115–121

X-ray images, 59

Yang, Chen Ning, 119
Yuan, Jada, 20

Zalinski, 31
Zalinski, Edmund, 31
Zenith "Space Command," 60
Zonta Society, 99